Green Technology and Sustainable Development

A Textbook for Modern Environmental Science

Ranjit K. Parmar, M.E. (Electrical-MSA)
Recipient of National Teachers Award 2024 (Hon'ble President of India) and
Pedagogy Innovation Award 2025 (Gujarat Technological University)

Department of Electrical Engineering,
Government Polytechnic Junagadh, Gujarat, India.

First Edition: 2025

Published by:
Ranjit K. Parmar,
50, Vrindavan Nagar, Near Madhuram Gate,
Junagadh, Gujarat, INDIA-362001

For feedback, suggestions, or to place orders, write us at:
parmar.ranjit@yahoo.com

Also available on Amazon, Kindle, Flipkart, GeM, and other platforms.

The net proceeds received from the sale of this book will be donated to support the education of underprivileged girls and to the PM CARES Fund.

ईशावास्यमिदं सर्वं यत्किञ्च जगत्यां जगत्।
तेन त्यक्तेन भुञ्जीथा मा गृधः कस्यस्विद्धनम्॥

Isha Upanishad, Verse 1

"Whatsoever exists in this ever-changing universe is gifted by God. Enjoy it, but do not covet what belongs to others."

Relevance to Green Technology and Sustainable Development

The very first verse of the Isha Upanishad guides us to live amicably with nature and its creatures. The resources on Mother Earth are not just meant for us; they need to be shared, respected, and preserved. It emphasises the judicious use of resources, avoiding overexploitation for the survival of our future generation. Technology should be such that it teaches the values of equity, protection, and affection for all. Thus, this verse teaches us the central theme of Green Technology and Sustainable Development.

Dedicated to

Navodaya Vidyalaya System

A source of inspiration, a second home for thousands of students, including myself. A system that shaped the thoughts and personality of students from rural and underprivileged backgrounds, empowered them to rise with equality and dream with confidence, and cultivated the seeds of humanity, discipline, and moral values.

Foreword

It is a matter of great pleasure for me to write a foreword for the book *"Green Technology and Sustainable Development"* authored by Prof. Ranjit K. Parmar. It covers the fundamental topics of the modern environment to serve as general knowledge for every common man due to global warming and climate change.

The book not only addresses renewable energy sources, but also a wide range of topics like EV, hydrogen energy, circular economy, green manufacturing, green transportation, e-waste, sustainable transportation, and sustainable agriculture, along with national and international climate policies. The book will motivate readers to adopt sustainable practices in their lives to reduce their carbon footprint from an individual perspective through engaging brainstorming activities, thinking corners, and real stories. Each chapter has been accompanied by lots of questions and MCQs to test the learning. I feel this book will serve to be a textbook for all undergraduate course students across India.

I noticed that the author has beautifully tried to interconnect various shlokas in our ancient literature of Sanskrit to showcase the harmonious living style of our ancestors.

I congratulate the author on his dedicated efforts for a remarkable contribution. I am sure this book will not remain just an academic reference but also a source of inspiration to showcase how technology can showcase how knowledge, innovation, and cultural wisdom can come together to address some of the most pressing challenges of our time in alignment with the UN Sustainable Development Goals.

Date: 29th August 2025

Prof. Pratapsinh Chauhan
Vice Chancellor
Bhakta Kavi Narsinh Mehta University,
Junagadh, Gujarat, India

Preface

Today, we are on the threshold of throwing ourselves into the deep, hot valley of global warming and climate change due to our actions of 'development,' which is damaging the delicate ecological balance. If we do not take urgent steps soon, then it will not be possible to reverse the actions initiated by nature via climate change. We are witnessing record-breaking scorching heatwaves, deluges, rains in the deserts, cyclones, and many unprecedented weather events devastating our homes and lives. Green Technology is the ultimate solution to combat the challenges of climate change, resource depletion, and socio-economic impacts of environmental degradation. It can show us the path to a sustainable future for the survival of our future generations on this planet.

This book, *Green Technology and Sustainable Development,* is a humble effort to encourage students across India to incorporate sustainability into their daily routines to conserve the precious natural resources. This book tries to present a comprehensive summary that explores the concept and need of green technologies; core renewable technologies like solar, wind, hydropower, geothermal, and tidal power; global and Indian policies; and practices for the sustainable transformation. This book not only explores the technical aspects but also provokes the reader to think about socio-economic inequality, human survival on this planet, and their role in fighting climate change. This book tries to give you a concise, comprehensive crux of green technologies, green practices, and sustainability in an easy and digestible format.

Green Technology and Sustainable Development are part of almost all undergraduate programs at universities, including Engineering, Science, Arts, Commerce, Management, Pharmacy, Agriculture, and others. The author has tried to keep this book as simple as possible, to suit almost all branches of various universities across India. The author has also tried to avoid mathematical equations, jargon of technical terms, and tough vocabulary, so that students may feel joy while reading. While reading the book, you may find many words and concepts repeating, but be assured, it is purposely done to remind you of our basic responsibility as humans towards our mother Earth and for our future generations.

Following the latest National Education Policy (NEP 2020), the author has introduced course outcomes and their mapping with PO. Throughout the book, Bloom's taxonomy has been followed to align with the international standard. A sufficient number of short and long answer questions, multiple choice questions, have also been added to evaluate the learning. In every chapter, readers will find brainstorming 'Thinking Corners' and thought-provoking 'Activities' along with motivational 'Real Stories' to engage them not just physically but also emotionally, promoting rigorous brainstorming by both mind and heart.

A Sanskrit shloka, related to the theme of the chapter, has been kept at the beginning of each chapter to appreciate the values of our rich culture and ancient practices to live amicably with nature, a central theme of Green Technology and Sustainability.

The author will be happy if this book can reduce carbon footprint at the individual level, even by a fraction.

Junagadh
30th August 2025

Ranjit K. Parmar

Acknowledgment

Authoring a book is always a time-demanding, tough, and brainstorming process that requires input from various individuals and resources to frame ideas and thoughts into words. I take this opportunity to express my gratitude to all those who have extended their cooperation, support, and encouragement while drafting this book.

I am thankful to Prof. Pratapsinh Chauhan, Vice Chancellor, BKNM University, for writing a foreword for this book; my friends Dr. Vinod Tejwani, Dr. Vimal Sonara, Dr. Balwant Makwana, Dr. Ramesh Padodara, and Er. Dharmesh Purohit for technically verifying the concepts and figures; to Dr. A. S. Pandya and Prof. Kishan Dave for their valuable suggestions to improve the book; and Prof. R. D. Vanara for helping with the figures and formatting of the book.

I would also like to acknowledge the Navodaya Vidyalaya System for shaping my thoughts towards sustainability. I also sincerely thank all staff of the Government Polytechnic Junagadh, and the Commissionerate of Technical Education, Gandhinagar, for their continued support.

I am also incredibly grateful to my parents, my wife Jigisha, and my daughter Rutva for their patience and motivation for the successful completion of this endeavour. My sincere thanks to Almighty God, who has always bestowed blessings on me.

For authoring this book, I have explored lots of online and offline resources, websites, search engines, AI tools, books, and consulted many colleagues, experts, and industrialists. I am thankful to all those creators whose information has directly or indirectly helped me to draft this book.

Junagadh
30th August 2025

Ranjit K. Parmar

Table of Contents

List of Appendices

Course Outcomes (COs)

Credits : 3(L-T-P: 3-0-0)   Prerequisites: Basic knowledge of environmental science.

At the end of this course, the readers will be able to:

CO	CO Statement	Related Chapters
CO1	Explain the basic concepts of Green Technology and sustainable development, given global challenges like climate change, pollution, and resource depletion.	1,2
CO2	Compare the core green technologies like solar, wind, hydropower, geothermal, and tidal energy with reference to their working principles, merits, demerits, and economic viability for sustainability.	3,4,5,6
CO3	Discuss the green infrastructures like green ICT, green buildings, smart cities, and their certifications.	7,8
CO4	Discuss innovative practices in agriculture, transportation for sustainability, using organic farming, EVs, and hydrogen fuel cells.	9,10
CO5	Apply circular economy and waste management practices to reduce the carbon footprint.	11
CO6	Interpret international treaties and initiatives, including UN SDGs.	12
CO7	Forecast the future trends of green technologies and their implications.	13

Mapping with PO

CO	Bloom's Level	Description	PO Mapping
CO1	K1-K2	Recall and understand definitions	PO1
CO_2	K3-K4	Apply and analyse renewable technology	PO2
CO3	K4-K5	Analyse and evaluate green infrastructure	PO6
CO4	K3-K5	Apply sustainable practices	PO3
CO5	K3-K6	Apply and create sustainable models	PO7
CO6	K4	Analyse global policies	PO11
CO7	K5-K6	Evaluate and predict innovation trends	PO12

CO = Course Outcomes; PO = Program Outcomes; K1–K6 = Bloom's taxonomy cognitive level

वसुधैव कुटुम्बकम्

Verse 72, Chapter 6, Maha Upanishad

"The World is One Family"

This verse of the Maha Upanishad is engraved in the entrance hall of the Parliament of India, emphasizing the "deep roots of love and respect for all creatures of the world," treasured in ancient Indian literature.

Chapter 1

INTRODUCTION TO GREEN TECHNOLOGY

Learning Outcomes:

After reading this chapter, the learner will be able to:

1. Understand the concepts of Green Technology and Sustainable Development.
2. Differentiate between conventional models of development and their sustainable alternatives.
3. Discuss the global challenges of resource depletion, climate change, and the energy crisis.
4. Recognize the importance of Green Technology in fulfilling the UN Sustainable Development Goals (SDGs).
5. Analyse the role of policy frameworks and regulations while dealing with social, ethical, and environmental challenges in shaping the green initiatives.

1.1 Defining Green Technology and Sustainable Development

Green Technology does not have a direct reference to the colour green, but it symbolically relates to plants. We have seen that the colour green is associated with plants, which are the ultimate source of our food and shelter, and, ultimately, our life in a sustainable way. Nowadays, in a broader sense, it has been associated with sustainable technology, and it refers to clean technology, which includes innovations and practices that mitigate environmental harm and promote sustainable development [1]. It also aims to solve the burning issues of global warming, climate change, resource depletion, and pollution to create

Real Inspiration

Under the guidance of Sonam Wangchuk, students from the icy region of Ladakh built an entire passive solar design school. Even when the ambient temperature drops as low as -20°C, the classrooms stay warm and comfortable for studying. [2]

Source: Wangchuk, S. (TED Talk) SECMOL, Ladakh.

solutions that are ecologically sustainable, economically viable, and socially adaptable [20], [1].
We may summarise the key concepts of Green Technology as to reduce carbon footprint by reducing resource consumption, to reduce pollution and waste by giving priority to renewable energy sources replacing fossil fuels; to focus on energy efficiency in industries, homes, and transportation sector through various ways like adopting LED lighting, five star rated electrical appliances, smart transmission grid; to reduce pollution by carbon capturing; to adopt circular economy by accepting 5R principle of Refuse, Reduce, Reuse, Repurpose and Recycle which emphasizes on relinquishing "Use and Throw" culture; to achieve sustainable development by following the United Nations' Sustainable Development Goals (UN SDGs) [9], [1], [11].

Applications of Green Technology include (a) renewable energy generation through solar panels, wind turbines, geothermal plants, and hydropower plants, (b) non-polluting transportation through electric vehicles (EVs) and hydrogen fuel cells, (c) waste management through biodegradable materials, and waste to energy conversion, (d) constructing net-zero carbon buildings (e) pesticides and chemical fertilizer use is also drastically reduced [28]. Thus, we can see that Green Technology has widespread applications and is widely beneficial to ensure human existence on this planet.

Now, let us talk about another word, Sustainable development. Sustainable development refers to the concept of fulfilling the present generation's needs without endangering the ability of our future generations to fulfill their needs [21]. In other words, we may say that resource consumption should be equal to resource generation [1]. It focuses on social equality, social responsibility, and economic viability, seeking a balance between human progress and planetary health. The Three Pillars of Sustainability are: [1]

(1)Environmental Sustainability: Conservation of natural resources, preservation of the ecosystem by focusing on climate change, waste reduction, and renewable energy.

(2)Social Sustainability: To ensure equity, quality, and justice for all by focusing on human rights, poverty, education, and healthcare sectors.

(3)Economic Sustainability: To achieve long-term economic growth without overexploiting natural or social capital by focusing on the circular economy, ethical business practices, and green jobs.

Technology and development are highly interconnected. Economic and social development is possible with innovations in technology, but a cut-off is required to be drawn to ascertain that development does not overexploit or pollute our precious resources [8]. Here, once again, technology comes into the picture, and innovations should be such that they assist ecological growth rather than deplete it.

> ***Thinking Corner**: Can you point out any of our innovations that are entirely free from environmental pollution?*

1.2 The Importance of Sustainability as of today

For the survival of humans on this planet, it is essential to have sustainable development while keeping an eye on rapid industrialization, climate change, and resource depletion. It is necessary to prioritize sustainability over any development. Development is necessary, but it should not be at the cost of climate change, global warming, and over-exploitation of resources. We shall have to conserve our natural resources for our future generations and human survival [7]. Governments, businesses, schools, colleges, and society must adopt eco-friendly practices to ensure balanced development. Global environmental challenges like deforestation, forest fires, biodiversity loss, and social and economic inequality need to be addressed following the

United Nations' Sustainable Development Goals (UN SDGs) [1]. Innovative technology may play a pivotal and crucial role in tackling these challenges. These SGGs are pillars of the framework for sustainable development. They are important because they encourage global collaboration between various stakeholders like individuals, societies, governments, and businesses while addressing the issues of climate change, global warming, poverty, inequality, and economic growth to ensure a better future for the next generations [24], [26].

Before proceeding further, let us see what the UN SDGs are: These are a set of seventeen global goals. They were adopted in 2015 [1]. They are:

1. No Poverty: Eradication of poverty and fair distribution of resources
2. Zero Hunger: Food for all by improved agriculture.
3. Good Health and Well-being: Disease prevention and well-being of all.
4. Quality Education: Lifelong learning through quality education.
5. Gender Equity: Empowerment of women to avoid gender discrimination.
6. Clean Water and Sanitation: Improved sanitation facilities and clean drinking water.
7. Affordable Clean Energy: Emphasis on renewable energy and improved efficiency of equipment and processes.
8. Decent Work and Economic Growth: Promotion of labour rights and employment with dignity..
9. Industry, Innovation, and Infrastructure: Industries with resilient infrastructure.
10. Reduced Inequalities
11. Sustainable Cities and Communities: Affordable housing.
12. Responsible Consumption and Production: Reduced resource utilization through the 5Rs principle..

13. Climate Action
14. Life Below Water: Protection of marine biodiversity by reducing plastic pollution in oceans.
15. Life on Land (Forest and Biodiversity Conservation): Protection of ecosystems, stopping desertification, and deforestation.
16. Peace, Justice, and Strong Institutions: Strengthening governance, the rule of law, and human rights.
17. Partnerships for the Goals: Global cooperation for sustainable development.

1.3 Global Challenges: Climate Change, Resource Depletion, and Pollution

Due to rapid industrialization and increased demand due to improved lifestyles, the 21st century is facing a threat to the stability of ecosystems and the existence of human beings at large. These challenges are not isolated. Problems may seem separate from each other, being in a variety of sectors, but they are interconnected with a complex web of our actions leading to a common risk. We shall have to focus on an integrated approach to combat these global challenges. The following is a list of a few calamities caused by climate change:

Year	**Climate-induced Calamity**
2015	Hurricane Patricia impacted Mexico's Pacific coast
2017	Hurricane Harvey, Irma, and Maria in Texas, Florida, and Puerto Rico
2017	California Wildfires
2019	Australian Bushfires
2020	East Africa Floods
2020	Cyclone Amphan affecting India and Bangladesh
2021	European Floods Affecting Germany and Belgium
2021	North American Heatwave
2022	Pakistan Floods
2022	South African Floods

2023	Global Wildfires Affecting Southern Europe, North America and Australia
2023	Typhoon Mawar impacted the Philippines
2024	Spanish Floods
2024	California Wildfire in the USA

Table 1.1 List of climate-induced calamities in the last decade

1. **Climate Change:** The Global average temperature has been increasing at a rapid rate since the end of the Industrial Revolution. The root cause of global warming is greenhouse gases (GHGs) like CO_2 and CH_4 because of their ability to retain heat [25]. They are produced mostly due to the burning of fossil fuels in industries and vehicles. It has also accelerated due to deforestation and mining. Forest fires are becoming increasingly common. A recent example is the wildfire of January 2025 in Los Angeles, California, USA, which killed more than 30 people, evacuated more than 2,00,000 people, destroyed more than 18,000 houses, and burned nearly 233 km^2 (roughly 57,500 acres) of forest. Due to climate change, we are facing not just wildfires but also hurricanes, floods, and droughts. Sea levels are constantly increasing, which has threatened the existence of several coastal cities like Miami, Jakarta and Mumbai, and a few low-lying island nations like Tuvalu. Global warming is responsible for the melting of Arctic ice, leading to the release of billions of tons of trapped carbon dioxide, which further increases global warming. We are trapping ourselves in a vicious cycle of global warming. Developed nations have contributed, and are continuing to contribute significantly more GHGs in comparison to developing nations. Figure 1.1 clearly depicts the saga of ever-increasing temperature post-industrialisation with the increasing CO_2 level. At present, China is the largest emitter of CO_2, while the USA has the highest cumulative CO_2 emissions due to its long industrial history. Refer to the following table for more details on CO_2 emissions:

Rank	Country	CO_2 Emissions (Metric Tons)	Percentage of Global Emissions	Per Capita Emissions
1	China	12,667,428,430	32.88%	8.89
2	United States	4,853,780,240	12.60%	14.21
3	India	2,693,034,100	6.99%	1.89
4	Russia	1,909,039,310	4.96%	13.11
5	Japan	1,082,645,430	2.81%	8.66
6	Indonesia	729,000,000	1.80%	2.48
7	Iran	686,415,730	1.78%	7.67
8	Germany	673,595,260	1.75%	8.01
9	South Korea	635,502,970	1.65%	12.27
10	Saudi Arabia	607,907,500	1.58%	18.89

Table 1.2 Country-wise CO_2 Emissions based on 2022 data.
(Source: Global Carbon Project 2023)

2. Deforestation and Biodiversity loss: We are losing our forests at the rate of approximately 10 million hectares per year (2015-2022 average). Forests are of utmost importance for biodiversity. As per the *IPBES 2019* report, one million species are at the edge of extinction due to human activities [26]. Major drivers behind deforestation are agriculture and blind industrialisation in the periphery of forests. 80% of Amazon deforestation may be linked to cattle ranching and shifting farming practices. Another example may be cited from Indonesia and Malaysia, where palm oil plantations have been favoured at the cost of forests. The paper industry also has a huge demand, which is solely responsible for tree cutting. Higher demand for timber has also directly contributed to Illegal logging, leading to deforestation. The reduced trees and jungle areas can turn forests into deserts, ultimately releasing billions of tons of trapped CO_2 [24].

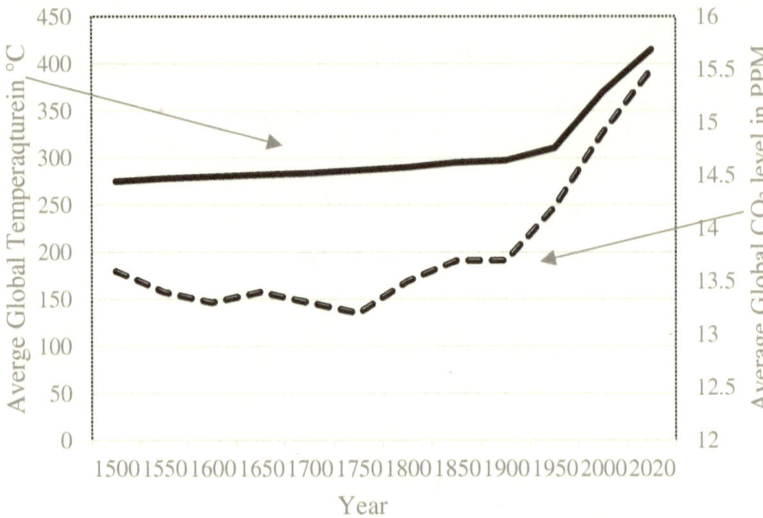

Figure 1.1 A graph showing a rise in temperature vs CO_2 emissions.
(Source for data: https://data.giss.nasa.gov)

3. Resource Depletion: According to BP's 2023 statistical review, it is estimated that present known oil reserves may hardly last the next 50 years if the present rate of consumption is maintained. Certain industrially important minerals like Lithium, Cobalt, and rare Earth elements have been in remarkably high demand for the last couple of years due to their importance in EVs and other technologies. Underground water is being depleted at three times the rate of its replenishment. 2.3 billion people are facing a drinking water crisis. 33% of global soil is degraded due to multiple reasons like soil erosion, deforestation, and pesticides, leading to food scarcity. According to the UNEP 2021 report, 11 million tons of plastic waste are dumped in the ocean every year, totalling 200 million tons [30]. This plastic breaks into microplastic granules. Through fish, they are now found even in humans, affecting human health. As we are more inclined towards gadgets, an annual E-waste of more than 53 million tons is produced. Out of which, a very meagre amount, 22.3%, is recycled. (*UN Global E-Waste Monitor 2023 report*)

2. **Pollution:** In simple words, we may define pollution as the contamination of air, water, or soil. Its concept may even be extended to space and sound. Pollution has become a prominent threat to human and planetary health. It is due to human activities and often a result of the accumulation of harmful activities over decades. Air is polluted through the combustion of fossil fuels in vehicles, industries, wildfires, and power plants. It is linked to respiratory diseases. According to the World Health Organisation (WHO), it causes nearly 7 million deaths per year [29]. Here, it is interesting to recall the forced closure of schools and cancellation of flights due to excessive pollution in Delhi, India. Common gaseous pollutants are carbon monoxide, sulphur dioxide, CFCs, and nitrogen oxides. Particulate matter or fine dust of size PM_{10} to $PM_{2.5}$ is also considered a pollutant. Water pollution is the contamination of water bodies like rivers, lakes, oceans, and water reservoirs, including groundwater. These water bodies may be contaminated by sewage discharges, industrial wastes, chemical wastes, oil spills, plastic, pesticides, pathogens, pharmaceutical pollutants, and others. These pollutants directly affect the environment as well as human and animal health. Noise pollution and Light pollution also disrupt our ecosystem. The following table shows various pollutants, their sources, and their health-related issues. Overexploitation of Earth's resources due to population growth and profit-driven extraction of minerals (fuelled by rapid industrialisation) has brought the ecosystem to the verge of brink. Even today, more than 80% of global energy

comes from fossil fuel-based sources like oil, coal, or gas. Mineral extraction has almost tripled since 1970.

Sr.	Pollutant	Severity	Sources	Health Issues
1	Particulate Matter (PM2.5 and PM10)	Very High	Vehicle emissions, industrial processes, construction, biomass burning	Respiratory diseases, lung cancer
2	Carbon Monoxide (CO)	High	Incomplete combustion of fossil fuels, vehicle exhaust, and industrial processes	Reduced oxygen delivery to organs, dizziness, headaches, and fatal at high exposure
3	Sulphur Dioxide (SO_2)	High	Burning of coal and oil, industrial emissions, and volcanic eruptions	Respiratory issues, asthma aggravation, lung diseases, and acid rain
4	Nitrogen Oxides (NOx)	High	Vehicle emissions, power plants, and industrial activities	Respiratory infections
5	Ozone (O_3) [Ground-Level Ozone]	Moderate to High	Reaction of NOx and VOCs under sunlight	Respiratory issues, chest pain, lung inflammation
6	Lead (Pb)	High (Very harmful to children)	Leaded gasoline (banned in most places), industrial emissions, lead-based paints	Neurological damage, developmental disorders,

				and kidney damage
7	Volatile Organic Compounds (VOCs)	Moderate to High	Paints, solvents, vehicle emissions, and industrial activities	Eye irritation, headaches, liver and kidney damage
8	Mercury (Hg)	Very High (Neurotoxic)	Coal combustion, mining, and industrial waste	Neurological disorders and kidney damage, Contamination of aquatic ecosystems
9	Black Carbon	Very High	Diesel engines, biomass burning, and industrial pollution	Lung and heart diseases
10	Ammonia (NH_3)	Moderate	Agricultural activities, fertilisers, and livestock waste	Respiratory problems

Table 1.3 Pollutants, their sources, and their health-related issues

We are consuming at a much faster rate than we are replenishing, leading to their depletion. This may lead to resource wars in the future and other socio-economic impacts and inequality. A few examples are worth recalling here. China mines 60% of its Rare Earth Minerals (REEs) for smartphones and other electronic gadgets or EVs, leading to toxic waste ponds that contaminate farmland. The Aral Sea shrank by 90% due to the diversion of rivers, leading to the destruction of fisheries in peripheral coastal areas. Fast fashion is also leaving big footprints; for example, the production of just one cotton T-shirt may require nearly 2,700 litres of water, amazing? But it is true! [51]. To tackle all these challenges, circular economy-based models are required to be adopted, which emphasise the reduce, reuse, and recycle concept. The transition to renewable energy should be prioritised. Crop rotation and focus on soil health should be given. Certain technical innovations may also help; for example, mycelium-based materials and lab-grown diamonds reduce mining demand. AI should be used to improve efficiency; for example, Google's DeepMind has slashed the cooling requirement for data centres by 40%. Briefly, innovative methods must be promoted and should be used for sustainable development.[7], [25].

1.4 An intersection of Green Technology and sustainable development

So far, from the previous discussion, it is clear that Green Technology is very much necessary for our future generations and even our survival, but at the same time, it is also very necessary to investigate the cost of the application of Green Technology. There are several challenges in implementing Green Technology.

(a) **High Cost:** At present, green technologies are much costlier than traditional technology, so they limit their access to developing countries.

(b) **Public awareness:** People are not aware of the long-term benefits of Green Technology.

(c) **Difficult financing:** It is difficult for small companies and underdeveloped countries to get finance for the implementation of Green Technology.

(d) **Higher R&D Cost:** The inventing company always tries to get maximum profit for its research and development efforts invested in the innovative technology, so in the initial years, every technology used to be much costlier, and Green Technology is much costlier at present. For example, an electric car costs almost double the price of a traditional IC engine car [23], [27].

To summarise, we may say that it is essential to make green technologies affordable for everyone [12], [19] by making Green Technology economically viable.

1.5 The Triple Bottom Line: People, Planet, and Profit

In the business world, the concept of the triple bottom line (TBL) is important. It addresses three key areas: (1) social responsibility (people), (2) environmental stability (planet), and (3) economic probability (profit). It was first introduced by *John Elkington* in 1994, who proposed that companies should not merely focus on finance and profit, but that it is a moral duty of all companies to consider the social and environmental impacts of their actions. The first pillar, people, or social responsibility, includes human rights, community development, and Product Safety. The second pillar, which is a planet or environmental stability, includes resource depletion, pollution, biodiversity loss, and climate change, while the third pillar, which is profit or economic viability, includes revenue growth, profitability, and return on investment. Adopting a TBL approach, we can provide certain benefits to society, workers, and businesses in the long term. For example, enhanced reputation or brand value, reduced regulatory risks, and improved employee engagement and retention. However, there are certain difficulties in measuring it, due to unawareness among the employees and people around industries, about their rights; conflicts between environmental and economic objectives; lack of a standardised framework for TBL; and false reporting of social and environmental impacts.[6],[16]

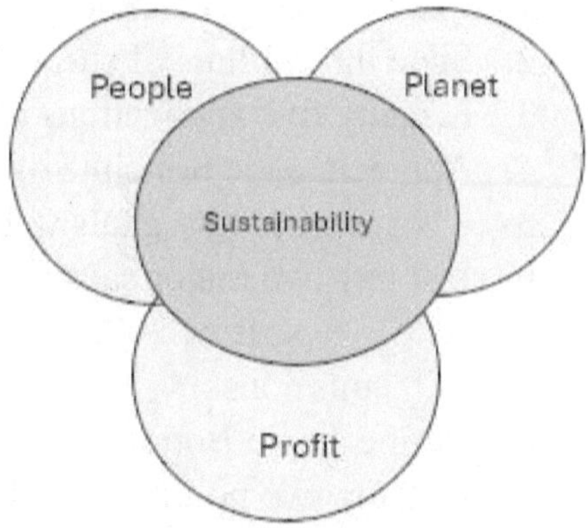

Figure 1.2 Interlinking of Three Pillars of Sustainability

In a concluding remark, we must add that green technologies should be affordable for everyone, including developing countries, and at an individual level. Governments, companies, and organizations should actively promote and implement non-polluting renewable energy sources and green technologies. Environmental and social responsibilities should be given priority over profit [1], [12], [19].

> ***Thinking Corner**: Suppose you are the owner of a company. How much share of your profit will you keep reserved for the benefits of your employees?*

> ***Activity:** Search the word CSR on internet. List various activities done by companies for the people.*

-: EXERCISE: -

Section A: Short Answer Questions

1. Define Green Technology.
2. State three pillars of sustainability.
3. List any five applications of green technologies.
4. Appreciate the principle of 5R in Green Technology.
5. UN sustainability goals were adopted in ..... year.
6. List any two major causes of climate change.
7. As per reports of 2022, China contributed .... % of global CO_2 emissions.
8. Define Triple Bottom Line.
9. List any two pollutants and health issues caused by them.
10. Recall steps in our daily life to reduce our carbon footprint.

Section B: Long Answer Questions

1. Justify the need for Green Technology.
2. State the key concepts of green technologies and their importance for sustainability.
3. Describe the applications of Green Technology in different sectors.

4. Explain how sustainable development is important for human existence.
5. Discuss the role of green innovations to support sustainable development.
6. List any three UN SDGs.
7. Discuss the global challenges: pollution and climate change.
8. Being Indian, list the challenges in the implementation of green technologies in a developing country like India.
9. Explain the role of forests in sustainability.
10. Briefly discuss the concept of the Triple Bottom Line to promote sustainable development.

Section C: Multiple Choice Q and A

1. .... is an example of Social Sustainability?
   A. To promote organic fertilizer
   B. To ensure access to quality education
   C. Carbon capture technology
   D. To increase mining capacity

2. The circular economy is different from traditional models as ......
   A. It relies only on government subsidies.
   B. It promotes one-time use of resources.
   C. It emphasizes reuse and recycling.
   D. It encourages maximum extraction.
3. .... Gas is primarily responsible for trapping heat.
   A. Hydrogen B. Methane
   C. Oxygen D. Neon
4. ....health hazard is linked with PM2.5.
   A. Teeth decay B. Leg aching
   B. C. Respiratory diseases D. Kidney failure
5. The best way to reduce urban air pollution is
   A. More Cars B. Planting more trees
   C. Construction D. Building dams.
6. Cyclone Amphan was the reason for:
   A. Deluge in India
   B. Cyclone in Singapore

C. Devastation in India and Bangladesh
D. Wind waves in Japan

7. A major factor in deforestation is:
   A. River erosion
   B. Industrialization near forests
   C. Solar energy expansion
   D. Urbanization
8. Which chemical pollutant causes acid rain?
   A. COx B. NO_x C. SO_2 D. H_2O
9. A practical example of industrial symbiosis is:
   A. Plastic recycling in homes
   B. Using heat from power plants in nearby homes
   C. Selling batteries to consumers
   D. Wind turbine energy export
10. .... had coined the concept of 'Triple Bottom Line'.
    A. Sunita Narayan B. John Elkington,
    C. Albert MacHall D. Chetan Solanki
11. Microplastics:
    A. Improves kidney health.
    B. Improves Ocean life
    C. Affects human health through fish.
    D. Reduces soil salinity.
12. Fast fashion is not environmentally friendly because:
    A. It encourages climate resilience
    B. It supports cottage industries
    C. It uses excessive water and energy
    D. It promotes traditional art
13. .... practice supports a green lifestyle.
    A. One-time use plastic B. Using LED bulbs
    C. Frequent car rides D. Buying bottled water daily
14. The Aral Sea shrank due to:
    A. Rising sea levels B. Cyclones
    C. River diversion for irrigation D. Volcanic activity
15. .... country has the highest per capita CO_2 emission.
    A. India B. Japan
    C. USA D. Saudi Arabia

16. Google DeepMind used ...... technology to reduce cooling energy.
    A VR
    B. QC
    C. AI
    D. BC

17. The .... gas is classified as a neurotoxin.
    A. Carbon Dioxide
    B. Nitrogen
    C. Mercury
    D. Ozone

18. The 5R principle is a part of:
    A. Greenhouse strategy
    B. Linear economy
    C. Circular economy
    D. Renewable policy

19. Which of these shows a commitment to intergenerational equity?
    A. Fast food consumption
    B. Space tourism
    C. Sustainable development
    D. Fossil fuel exploration

20. The UN SDG 'Climate Action' is goal number:
    A. 17
    B. 10
    C. 13
    D. 5

21. The average annual loss of forest is … million hectares.
    A. 1
    B. 5
    C. 10
    D. 15

22. A major reason for climate change is:
    A. Organic farming
    B. Fossil fuel burning
    C. Wind energy
    D. Plantation programs.

23. .... is not a pollutant.
    A. Particulate matter
    B. SO_2
    C. Black carbon
    D. O_2

24. 'No Poverty' is included in .... UN SDG goal?
    A. 1
    B. 7
    C. 4
    D. 11

25. Which principle is emphasized by the quote "Mata Bhumi Putro Aham Prithivyah"?
    A. Capitalism
    B. Ownership of land
    C. Ecological responsibility

D. Political independence

26. UN SDG 12 deals with what?
    A. Gender equality B. Clean water
    C. Sustainable consumption D. Hunger eradication
27. The main concern of noise pollution is:
    A. Mind relaxation B. Mental stress
    C. Ear exercise D. UV exposure
28. .... is an example of policy-level climate action.
    A. Tree planting B. Paris Agreement
    C. Household composting D. Using bicycles
29. The soil fertility can be retained by sustainable practice.
    A. Medicines B. Synthetic Fertilizers
    C. Mango Plantation D. Crop rotation
30. A country to become the first carbon-neutral capital by 2025?
    A. Tokyo B. New Delhi
    C. Copenhagen D. Amsterdam

Answer Keys:

1	2	3	4	5	6	7	8	9	10	11	12	13	14	15
B	C	B	C	B	C	B	C	B	B	C	C	B	C	C
16	17	18	19	20	21	22	23	24	25	26	27	28	29	30
C	C	C	C	C	C	B	D	A	C	C	B	B	D	C

"माता भूमिः पुत्रोऽहं पृथिव्याः"

अथर्ववेद 12.1.12

"Earth is my Mother, and I am her child."

Atharvaveda, Book no. 12 (Bhoomi-sukta), Hymn 1, Verse 12

Chapter 2

FOUNDATION OF GREEN TECHNOLOGY

Learning Outcomes:

After reading this chapter, the learner will be able to:

1. Trace the evolution of green technologies from ancient non-sustainable practices.
2. Explain the principle of the Green Economy and its relevance in national and international policies.
3. Compare conventional industrial technologies with green technologies.
4. Describe the turning point treaties like the Earth Summit, the Kyoto Protocol, and the Paris Agreement.
5. Evaluate industrial growth, consumer behaviour, and global trade to shape green innovations.

.

2.1 The Evolution of Green Technology

It has been said in ancient Sanskrit text, **वसुधैव कुटुम्बकम्** (*Vasudhaiv Kutumbakam*), meaning all members on this planet are part of one family. Mother Earth has already bestowed sufficient resources for our needs, but not for our greed. Initial civilizations lived in harmony with nature, but human greed for wealth has overexploited nature, leading to climate change and global warming, which has led to climatic catastrophes. This has forced man to think about Green Technology. It is often referred to as clean energy. This section explores the trajectory of Green

Real Inspiration

Rehan, living in a village near Varanasi, built a solar lantern with the help of broken toys and e-waste after watching videos in cyber café. This solar lantern is helpful for her weaver mother to work more to fulfil the dream of her son to become a doctor.

Source: The Better India, 2022

Technology from its infancy to today's sophisticated innovations of the modern era.

Early Attempts at Sustainable Development:

Green Technology may be traced even in ancient civilizations, where people were living in harmony with nature. The following are a few examples:[21]

1. Water Management Systems: Ancient Indus civilization and Mesopotamia developed a sophisticated irrigation system to maximize the use of water in agriculture. For instance, the people of Persia (modern-day Iran) utilized the **Qanat System** to transport water from water-rich areas to arid regions through channels to reduce evaporation and simultaneously ensure a continuous water supply. **Rani ki Vav,** a Step Well in Patan, Gujarat, India *(UNESCO World Heritage Centre, 2023),* is not just a marvellous example of architecture but also showcases an efficient water storage system.
2. Passive Solar Design: To have the maximum solar heat inside the house, Greeks and Romans used to design their buildings with windows facing south, relinquishing the need for external heat sources. The Trombe wall concept was used in ancient China to capture maximum solar heat. It is still used in passive solar architecture. In the Roman hypocaust system, heated air from furnaces was used to warm rooms; it is like modern-day central heating. Thus, they were smart enough to conserve and efficiently use energy sources while reducing waste.
3. **Wind and Water Mills**: In your textbooks for class 10th, you would have heard about the history of windmills and water mills for various purposes. For example, the windmills of Persia (500-900 AD) were utilized to pump water and grind grains. The early industrialization period in Europe gained momentum through water wheels powering mills for textiles, lumber, and metalworking. These were the early

practical applications of renewable energy, where wind was used as an energy source.

These innovations lacked the scientific rigor of modern technology, but they were abreast of nature, and they had a deep sense of responsibility towards the ecosystem. These early systems have become role models to guide us in the present Green Technology.

> ***Thinking Corner****: In returning to nature-friendly living, we follow the footsteps of our ancestors. So, who truly understood nature better—us or them?*

The Green Revolution and Beyond

The Green Revolution of the mid-20th century proved a turning point in the evolution of Green Technology. It embarked upon significant advancements in farming to fulfil the goal of increased food demand due to an increased population. Key development includes:

1. **High-Yield Crops:** With the advent of genetics, scientists developed improved quality crops, which were able to give more food grains for less land, reducing the need for farm expansion. **Norman Borlaug**, who is commonly known as the **father of the Green Revolution**, developed wheat varieties that were dwarf, disease-resistant, and yielded higher wheat grains. This revolution boosted food production in countries like India and Mexico. Such inventions had averted famine in many parts of the world, but they also reduced biodiversity, as farmers had started to cultivate only modified, yielding crop varieties.
2. **Pesticides and Fertilizers:** Each coin has two sides. While chemical pesticides help to kill the pathogens producing disease in crops, and chemical fertilizers help farmers boost the crop yield, at the same time, pesticides and chemical fertilizers have side effects of degrading the soil quality, water pollution, and entering our food cycle. These chemicals are also contaminating our water bodies, leading to an imbalance in the aquatic ecosystem. The Dead Zone in

the Gulf of Mexico was created due to agricultural runoff from the Mississippi River and is a prominent example to showcase the consequences of intensive farming using pesticides and chemical fertilizers [24].

3. **Irrigation Technologies**: The Latest development in the field of irrigation technology has made it possible to extract water from deep in the crust. This has resulted in the over-extraction of groundwater, thus, depletion of the water table. For example, the Punjab state of India has been famous worldwide for wheat production, but overexploitation of groundwater is a red alarm for maintaining sustainable agriculture. Excessive irrigation for cotton farming has led to the almost disappearance of one of the world's largest lakes, the Aral Sea, and is also an alarm to maintain sustainable water management [30].

The green revolution has surely assured us of the higher production of grains and other crops to fulfill the demand of the increased population, but has also shown a negative impact of overexploitation of limited resources and contamination of water, soil, and air. It has also led to the entry of chemicals and pesticides into our food web, resulting in cancer and other diseases. A more comprehensive approach is required to balance agriculture and the environment to maintain sustainability.

Milestones in Green Technology

Research and scientific advancements, and environmental awareness, played a very pivotal role in the evolution of Green Technology. Global warming, climate change, and climate-induced catastrophes have mandated man to rethink the blind race of unsustainable development. The last few decades have witnessed significant milestones in the development of Green Technology. Key milestones are:

- **1970s: Environmental Movement:** With the establishment of the U.S. Environmental Protection Policy in 1970, the global movement for environmental awareness started. In 1972, the Club of Rome highlighted the consequences of unchecked economic growth in a report, *The Limits to Growth.* The oil crisis of 1970 promoted investments in alternative energy sources rather than fossil fuels. *The Clean Air Act* and the *Clean Water Act* set the regulatory framework for the first time in human history for the promotion of Green Technology.

- **1980s: Beginning of Renewable Energy:** Sole dependency on oil was challenged for the first time by the 1970s oil crisis. It spurred investment in alternative renewable energy sources like wind, solar, and hydropower. The first commercial solar panels and wind turbines were developed during this decade. Thousands of wind turbines were installed during the California Wind Rush, laying the foundation of modern wind energy. On the other hand, the Solar Energy Research Institute (SERI), which was established in this decade, played a particularly significant role in the development of solar power generation technology through various methods.
- **1990s: Climate Change Awareness:** A Significant step towards international collaboration towards climate action was taken through the *Kyoto Protocol* in 1997, binding developed countries to set emission reduction targets. The International Panel on Climate Change (IPCC) was established in 1988 to provide scientific assessments of climate change. The United Nations Framework Convention on Climate Change (UNFCCC) succeeded in bringing global attention to the need for decarbonization and sustainable development.
- **2000s: Energy efficiency through Green Buildings:** For the first time, through the Leadership in Energy and Environmental Design (LEED), a certification system was introduced to promote sustainable building designs to conserve energy. Energy conservation was further assisted by domestic energy-efficient appliances and smart grid

technology. People were motivated to buy energy-efficient appliances through programs like Energy Star. It was launched in 1992 and soon became the global standard. Germany developed new benchmarks for ultra-low-energy buildings through the Passivhaus.

- **2010s: Digital Transformation and Circular Economy:** A historical global commitment to reduce the global temperature rise below 2°C was crafted in this decade, known as the *Paris Agreement* in 2015. The concept of the 5Rs to reduce waste, recycle, and efficiently use resources gained momentum through the circular economy. Few companies, like Patagonia and Interface, accepted the challenge and implemented the model of the circular economy, promoting environmental sustainability. Resource management became better through the integration of advanced technologies like the Internet of Things (IoT) and Artificial Intelligence (AI).
- **2020s: Net-Zero Commitments:** Seeing the ill effects of climate change globally, many governments and corporations committed to net-zero emissions by 2050. The concept of a zero-carbon footprint became popular in this decade, spreading awareness and the need for it. The COVID-19 pandemic taught a lesson to humanity, leading to the acceleration of the adoption of Green Technology and sustainable development. For instance, the *European Green Deal* has set a goal to make the climate neutral by 2050 by prioritizing investments in Green Technology [10], [7].

Sustainability and Green Economy

Achieving a sustainable future requires a paradigm shift from resource-intensive growth models to green economic systems. The ever-increasing demand from the large population requires a balance in resource depletion, along with recycling and proper waste management. This section explores the models of sustainability and

> ***Activity**: Locate a map of Bhutan on the internet. List a few practices adopted by Bhutan for the well-being of their people. You may use online resources to get a feel of being in Bhutan! Try to feel the heart-touching beauty of Mt. Himalaya.*

green economy for the required growth while conserving the environment.

Economic Growth Vs Sustainability Balance

You might have heard of an innovative practice adopted by a small nation, Bhutan, located in the lap of Mount Himalaya, which emphasizes Gross National Happiness (GNH) rather than Gross Domestic Product (GDP) as the measure of overall well-being and happiness. Our traditional models use GDP to measure progress, neglecting environmental and social well-being. Now, the rigorous over-exploitation of finite resources and the impacts of climate change have shown the limitations of this model. Sustainable development ensures judicious use of resources to conserve for future generations, too. Figure 2.1 shows that resource depletions are much higher than resource replenishments due to loss and waste.

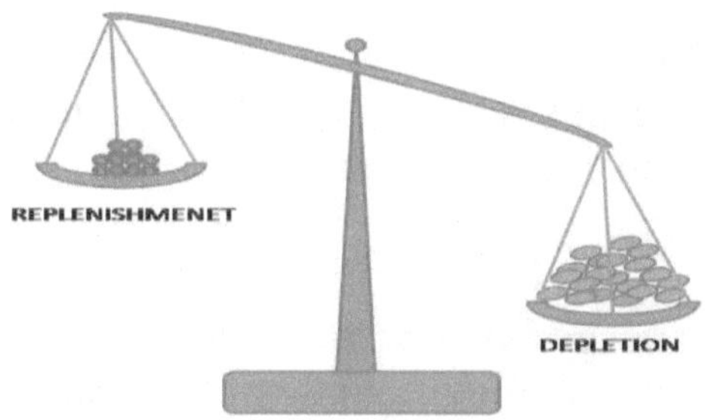

Figure 2.1 An illustration to show: depletion is much higher than replenishment

The following points are important from the perspective of economic growth and sustainability:

1. Decouple Economic Growth from Resource Use: This principle emphasizes achieving growth without environmental impact while reducing resource consumption through efficiency and innovation. For example. Denmark is now generating more than 50% of its electricity from wind power; thus, it has decoupled its economic growth from depleting resources by investing in renewable energy sources.
2. Carbon Taxation: This principle incorporates the taxation for environmental degradation, such as carbon pricing and pollution taxes. For example, the European Union Emissions Trading System (EUETS) and British Columbia's carbon tax incentivize emission reductions without harming economic growth.
3. Inclusive Growth: This principle ensures the sharing of economic growth, addressing social inequalities to enhance the quality of life and living standards for all. UN SDGs discussed in the previous chapter reflect the importance of inclusive growth, whether it is a developed or a developing nation.

Green Economic Models

The green economy comprises environmental sustainability and social equity, along with economic growth. It includes the following components:

1. **Renewable Energy Investments:** The primary focus of the green economy is shifting from fossil fuel energy sources to renewable energy sources such as wind, solar, and hydropower for the reduction of greenhouse gases. For example, Germany has introduced the *Energiewende* (Energy transition) policy to shift to low-carbon energy by 2050. To achieve this target, the country is heavily investing in solar and wind power plants. India is rapidly expanding its renewable energy sector through both public and private

partnerships and various incentives. Figure 2.2 shows the growth of the renewable energy sector in India.

2. Green Infrastructure: Sustainable urban planning is required through the strengthening of the public transportation system, incentives for green buildings, and smart cities. A global network of C40 cities is committed to planning for sustainable development. Copenhagen is planning to become the world's first carbon-neutral capital by 2025 through green infrastructure and investments in the renewable energy sector [10].
3. Practising Circular Economy: Maximum utilization of resources through improved efficiency appliances, waste reduction, and implementation of the 5R concept fulfils the circular economy principle.[10],[18] Many countries are adopting a circular economy. For instance, the Netherlands has planned to achieve a fully circular economy by 2050.

The transition to a green economy is not an instantaneous task, but it requires coordinated efforts from governments, business houses, societies, and individuals. Government policies in taxation and subsidies for adopting green energy play a vital role. For example, the Indian Government is heavily subsidizing the installation of solar power plants through the PM Suryaghar Yojana, which aims to provide free electricity to one core household in India through PV-based solar rooftop power plants. Figure 2.2 supports the fact that India is on the fast track to adopting renewable energy to meet its ambitious targets [12].

Measuring Sustainability Success: Certain organisations, like the Global Reporting Initiative (GRI) [1], provide a framework to measure the environmental, social, and governance (ESG) outcome through various parameters like waste generation, greenhouse gas generation, energy consumption, and social impact. Many companies like Nestle and Unilever use this framework to track and report their performance on sustainability.

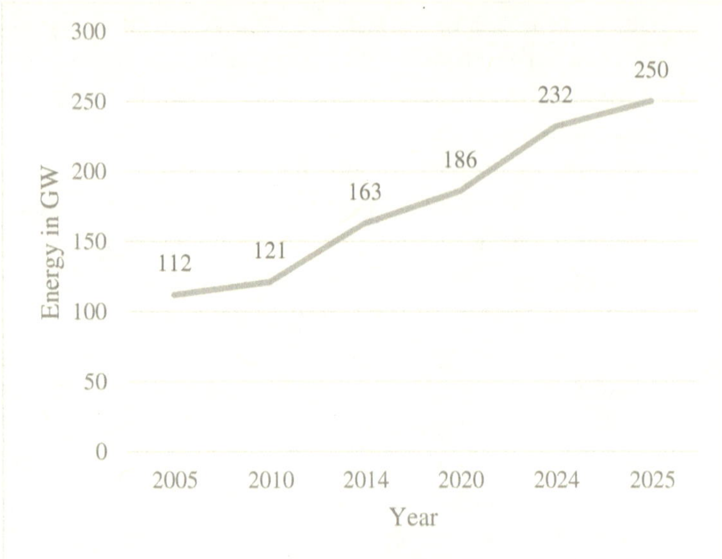

Figure 2.2 Cumulative Renewable Energy in the Last Two Decades in India
(source: https://mnre.gov.in/en/year-wise-achievement/ and IEA Data (2005–2025)

2.3 Technology to Promote Green Energy

The scientific principle behind Green Technology is to promote the efficient use of resources through sustainable practices. This section addresses various green technologies and decarbonization methods revolving around the circular economy.

- **Solar Energy:** The Sun is the ultimate source of energy for all. Solar energy can be converted to electrical energy through various means, like (a) photovoltaic cells (PV Cells) using semiconductor material and (b) concentrated thermal heating using mirrors. Many innovative technologies are being researched to increase the efficiency of solar cells and reduce the price of electricity per unit of electricity.
- **Wind Energy:** The kinetic energy of wind is used to generate electricity. Offshore and marine wind farms play a

significant role in renewable energy generation. Coastal states like Gujarat and Tamil Nadu are leading the wind energy sector. Improved materials and contemporary designs for wind turbines have increased the overall efficiency of windmills.

- **Hydropower:** It uses the potential energy of water to generate electricity. Water at higher heights is passed through water turbines for electricity generation. Small-scale hydropower and pumped storage systems provide options to meet the peak demand and stability of the power system. The largest hydropower projects in India include the Tehri Hydropower complex, having a capacity of 2400 MW, followed by the Koyna Hydroelectric project with a capacity of 1960 MW. At a global level, the largest hydropower project is the Three Gorges Dam (22,500 MW), followed by the Baihetan Dam (16,000 MW), both in China.
- **Geothermal Energy:** As we go deep inside the Earth's crust, we find a gradual increase in the temperature. This internal heat of the Earth is tapped by the geothermal system. It is used to generate electricity as well as to provide direct heating in colder regions. Iceland's Hellisheioi Power Station stands as the largest geothermal power plant, generating 303 MW of electricity and 133 MW of thermal energy [4].
- **Bioenergy:** This energy is derived from organic materials like plants, algae, and waste. Prominent traditional forms of bioenergy are biomass, biofuels, and biogas. However, advanced biofuels like cellulosic ethanol offer higher energy with a lower environmental impact.

These green technologies, along with a few others, will be discussed in detail in the next few chapters.

Decarbonization and Carbon Capturing

It refers to the reduction in emissions of greenhouse gas, mainly CO_2. The main source of CO_2 emissions is fossil fuel combustion.

The following strategies are used for decarbonization and carbon capture.

- **Energy Efficiency:** The Overall efficiency in buildings, transportation, and industries is increased through various practices to reduce the overall energy demand. The latest LED lighting and smart thermostats are examples of a few simple steps to reduce energy demand, thus reducing carbon emissions. For example, the Energy consumption of the Empire State Building was reduced by almost 38% through the retrofit project.
- **Electrification:** Wherever possible, replace the traditional fossil fuel-based technology, system, and equipment with electric alternatives. For example, Electric Vehicles.
- **Carbon Capture Storage (CCS):** This technology is used to capture CO_2 emitted from various industrial processes and power plants. This captured CO_2 is then transported and stored, preventing its release into the atmosphere. Tuticorin Carbon Capture Plant in Tamil Nadu, India, is capturing 60,000 tons of CO_2 per year [7]. This plant captures CO_2 from a fired boiler and converts it into sodium carbonate, that is, soda ash. The Gorgon CCS project in Australia captures 4 million tons of CO_2 per annum. The following is a summary of a few major CCS sites across the world.

Project Name	**Country**	**Capacity (MT CO_2/Year)**	**Industry**
Gorgon CCS	Australia	4.0	Bioethanol
Northern Lights CCS	Norway	1.5	Industrial CO_2 Storage
Petra Nova CCS	USA	1.4	Coal Power
Quest CCS	Canada	1.2	Hydrogen (Blue Hydrogen)

Illinois Industrial CCS	USA	1.0	Bioethanol
Sleipnir CCS	Norway	1.0	Natural Gas
Abu Dhabi CCS	UAE	0.8	Steel Industry

Table 2.1 Major CCS sites across the world.

- **Carbon Utilization:** This involves the conversion of CO_2 into various products, thus caging the emitted CO_2 to chemicals and bricks through processes like electrochemical reduction, mineralization, etc. Small steps from every individual can help reduce carbon utilization, like using paper or plastic waste in low-lying dump yards rather than burning.

- The following is a summary of a few major CCS sites in India.

Project Name	**State**	**Capacity (Tons CO_2/Year)**	**Industry**	**Operator**
Tuticorin Carbon Capture Plant	Tamilnadu	60,000	Chemical Industry (CCU)	Carbon Clean Solutions and Tuticorin Alkali
NTPC Vindhyachal CCS Pilot	Madhya Pradesh	Small-Scale	Coal Power (Pilot)	NTPC
Tata Steel CCUS Pilot	Jharkhand	Small-Scale	Steel Industry (Pilot)	Tata Steel
ONGC CCS for EOR (Planned)	Gujarat	Planned	Oil and Gas	ONGC

Table 2.2 Major CCS sites in India.

Circular Economy and Closed Loop System

As discussed earlier, it focuses on the 5Rs for sustainable development along with maintaining economic growth, which includes reducing, reusing, recycling, and repurposing. Thus, a system is closed in a way to maximize resource utilization. Its key components are:

- Long-lasting Design: Products should be designed such that they are more durable, upgradable, and easy to repair. Their spare parts should be easily available to maintain them in the market and operate for a longer period.
- Recycling: The need for virgin resources is greatly reduced if used products are recovered for recycling.
- Biodegradable Materials: Accumulation of plastics and other toxins can be reduced if we design products from biodegradable materials. They can be safely decomposed over a period without harming our environment. For example, now many companies are using biodegradable material, Polylactic Acid, a plastic-like material derived from corn starch, for packaging and disposable products.
- Industrial Symbiosis: Waste from one industry may be a resource for another industry, sometimes after a few modifications; but at large, it reduces the waste and impact on our environment and the need for virgin resources. For example, waste heat from power plants may be used to provide heat to nearby homes in colder regions. The Kalundborg Symbiosis in Denmark is one such example [1]. Thus, the circular economy results in the better utilization of resources and reduced impact on our environment while achieving sustainable growth. It is a closed-loop system that aligns with the principle of Green Technology.

In a concluding remark, we may say that Green Technology has evolved to address the challenges of climate change. By adapting the green economic models, we can achieve sustainable development that balances economic growth with environmental preservation.

***Thinking Corner**: Suppose your family is trapped on an island without electricity and electronic gadgets. What activities will you do to enjoy with your family?*

-: EXERCISE:-

Section A: Short Answer Questions

1. Briefly state how ancient Indian values were in alignment with *Green Technology*.
2. Discuss passive solar designs in ancient China.
3. Express your balanced views with regard to chemical fertilizers in the green revolution.
4. Name a global agreement to limit temperature rise.
5. Recall the concept of the 5Rs in the context of the circular economy.
6. Describe the basic principle of Green Technology.
7. List two leading states of India in wind energy generation.
8. State the full form of CCS along with its purpose.
9. Geothermal energy uses internal heat for energy generation.
10. Define circular economy and state its benefits.

Section B: Long Answer Questions

1. Compare the efforts of early civilizations to conserve water and energy with today's Green Technology.
2. List major achievements and environmental consequences of the Green Revolution.
3. Compare the traditional GDP-based model to measure economic growth with Bhutan's Gross National Happiness approach.
4. List key components of a green economy.
5. List five global examples of renewable energy investment, green infrastructure, and circular economy.
6. Discuss the role of natural forces in achieving sustainability.
7. Explain the role and impact of CCS in sustainability.

8. List challenges in the implementation of Green Technology.
9. Elaborate on the strategies of decarbonisation in reducing GHG emissions.
10. List technologies to promote green energy.

Section C: Multiple Choice Q and A

1. The ancient Sanskrit phrase "वसुधैवकुटुम्बकम्" (Vasudhaiv Kutumbakam) means...
   A) Technology must advance without limits
   B) All members of the planet are part of one family
   C) Earth is full of resources for everyone's greed
   D) Let us live peacefully and enjoy the food.
2. The ancient civilization had developed a sophisticated irrigation system to maximize water use in agriculture:
   A) Egyptian civilization B) Indus civilization
   C) Roman civilization D) Greek civilization
3. The purpose of the Qanat System used in Persia was:
   A) To generate electricity from wind
   B) To transport water from water-rich to arid areas
   C) To grind grains using waterpower
   D) To trap solar energy inside buildings

4. An ancient Indian architectural feature known for efficient water storage is:
   A) Step Well at Patan (Rani ki Vav) B) Rana ki Talavdi
   C) Navghan kuvo (Well of Junagadh) D) Adi-Kadi Vav
5. What concept did the Greeks and Romans use in building design to maximize solar heat?
   A) Trombe wall B) Windows facing south
   C) Water heating using Solar energy D) Sun curtains
6. In Persia, windmills were primarily used for:
   A) Cooling homes
   B) Pumping water, and grinding grains
   C) Heating homes
   D) Textile production
7. .... is known as the "father of the Green Revolution".

A) Norman Borlaug B) Swami Nathan
C) Rang Nathan D) Rachel Carson

8. The major negative impact of the Green Revolution's use of pesticides and fertilizers:
   A) Less food production
   B) Degradation of soil quality, and Water pollution
   C) Affordable food for all
   D) Increased water table
9. The .... was an environmental disaster due to agricultural runoff from pesticides.
   A) Water scarcity in America B) Dry Gulf of Hudson
   C) Evaporation of the Red Sea D) Dead Zone in the Gulf of Mexico
10. Over-extraction of groundwater in Punjab, India, led to:
    A) Reduced crop yield B) Water table depletion
    C) Excessive rainfall D) Increase in biodiversity
11. The consequences of unchecked economic growth were highlighted in the.............. global event in 1970.
    A) Kyoto Protocol
    B) Club of Rome's *Limits to Growth* report
    C) Paris Agreement
    D) California Wind Rush
12. In the 1980s, significant renewable energy development had occurred:
    A) First use of fossil fuels.
    B) Commercial solar panels and wind turbines
    C) Development of passive solar design
    D) Launch of the Paris Agreement
13. The .... aimed to reduce emissions.
    A) Earth Talks B) Kyoto Protocol
    C) Clean Air Act D) Doha Protocol
14. The LEED certification promotes:
    A) Chemical pesticide use
    B) Sustainable building designs for energy conservation
    C) Over-extraction of groundwater
    D) Windmill construction
15. A key global agreement to keep the global temperature rise below 2°C?
    A) Kyoto Protocol B) Paris Agreement

C) Club of Rome Report D) Clean Water Act

16. Bhutan's Gross National Happiness (GNH) differs from the traditional GDP in:
    A) GDP includes environmental sustainability
    B) GNH emphasizes overall well-being and happiness, including environmental and social factors
    C) GDP focuses on social equity and happiness for poor people.
    D) GNH measures economic output only
17. The ....generates over 50% of its electricity from wind power?
    A) UAE B) Denmark
    C) Singapore D) Taiwan
18. The carbon taxation is intended for:
    A) Increase economic growth by reducing taxes
    B) Penalize environmental degradation and incentivize emission reductions
    C) Replace renewable energy subsidies
    D) Subsidize fossil fuel industries
19. Choose a non-component of the green economy.
    A) Renewable energy investments B) Green infrastructure
    C) Circular economy practices D) Increased fossil fuel extraction
20. The Energiewende policy belongs to which country?
    A) India B) China
    C) Germany D) Australia
21. The .... scheme aims to provide free electricity through solar rooftop power plants in India.
    A) PM Ujjwala Yojana B) PM Suryaghar Yojana
    C) PM Ghar Ghar Bijli Yojana D) PM Free Bijali Yojana
22. Performance criteria for sustainability for the Global Reporting Initiative framework are based on:
    A) Energy generation B) ESG outcomes
    C) Forests D) Energy consumption only

23. A conversion of solar energy into electricity using a semiconductor material is due to
    A) Photovoltaic cells B) Solar turbines
    C) Photovoltaic cells D) Sun Mill

24. The principle behind carbon capture and storage (CCS) is:
    A) Preventing CO_2 emissions by storing them underground or converting them to other products
    B) Replacing fossil fuels with electric vehicles to reduce carbon emissions
    C) Increasing fossil fuel extraction efficiency to capture carbon.
    D) Burning waste materials and then landfilling them.
25. The CCS Tuticorin, a plant to capture carbon, is situated in:
    A) UP B) Tamil Nadu C) MP D) Kerala
26. The key feature of the circular economy is:
    A) Continuous use of virgin raw materials to get pure material
    B) Waste reduction and maximum utilization of resources through the 5Rs
    C) Increased fossil fuel usage
    D) Expansion of landfill sites to prevent pollution
27. Industrial symbiosis refers to:
    A) A treaty by industries to sell products at a higher price
    B) Waste from one industry is used as a resource by another industry
    C) Complete automation of industrial processes
    D) Isolation of industries to prevent pollution.
28. A biodegradable material derived from corn starch (used in packaging) is:
    A) Polylactic Acid B) Polyethylene
    C) Polystarch D) Polycorn maize
29. A method that can not reduce carbon emissions is:
    A) Energy efficiency improvements
    B) Electrification of fossil fuel systems
    C) Increased use of coal power plants
    D) Carbon capture and storage
30. "Decarbonization" aims to reduce:
    A) Noise pollution B) Soil erosion
    C) Greenhouse gas CO_2 emissions D) Air pollution

Answers Keys:

1	2	3	4	5	6	7	8	9	10	11	12	13	14	15
B	B	B	A	B	B	A	B	D	B	B	B	B	B	B
16	17	18	19	20	21	22	23	24	25	26	27	28	29	30
B	B	B	D	C	B	B	C	A	B	B	B	A	C	C

ॐ तत्सवितुर्वरेण्यं भर्गो देवस्य धीमहि।
धियो यो नः प्रचोदयात्॥

Rigveda 3.62.10

Let me meditate on the Sun God.

Give me a higher intellect and let the Sun God illuminate my mind.

Chapter 3

SOLAR ENERGY

Learning Outcomes:
After reading this chapter, the learner will be able to:
1. Explain the terms like solar radiation, insolation, and solar constants.
2. Describe the working of PV cells and differentiate their types.
3. Compare distinct types of solar systems, viz. PV, CSP, and Passive.
4. Analyse the integration of solar energy with the grid.
5. Evaluate the impact of solar energy on the environment and society.

3.1 Introduction to Solar Energy

Since the birth of the Earth, the Sun has remained the ultimate source of all energy for us. Solar Energy is derived from the Sun's light and heat through its radiation. It is the most abundant and clean source of energy. Seeing the threat of growing global warming, climate change, and the dearth of clean energy sources, solar energy is emerging as the leader among all renewable energy sources, as it does not produce GHG gases like fossil fuels, it is almost inexhaustible, and it is environmentally friendly. We receive an immense amount of energy from the sun. It is 10,000 times the world's energy requirement; it is almost 173,000 TW! If we can harness it through various means like PV cells, solar thermal systems, and other technologies, then the dependency on fossil fuels can be decreased.

Real Inspiration
With the help of NGO Gramin Vikas Trust, women from Barmer, Rajasthan shifted from manual sewing machines to solar-powered machines. Thus, productivity and profit increased, enabling girls to go to school.

Source: The UNDP India Report, 2020

Just close your eyes and appreciate the role of the Sun in our lives....!

> ***Thinking Corner**: We have developed a lot in many fields. Are we developed enough to live in the absence of the Sun? Think, what changes will be there in our life if the sun suddenly disappears from the sky?*

We shall discuss these things in detail in the following sections.

3.1.1 The Role of Solar Energy in Green Technology

Solar energy can play a significant role in delivering non-fossil fuel-based clean energy. It can reduce the carbon footprint, help to increase energy security, reduce pollution, and conserve water as compared to traditional power generation methods. As discussed in the earlier chapters, Green Technology emphasizes sustainable development through minimum environmental impact. Solar energy can supply electricity and heat, and can even help in the transportation sector through EVs. It can also help supply fresh water in arid regions through desalination plants. Its role extends beyond energy and environmental benefits. You will be amazed, but it also has the potential to create jobs and stimulate economic growth. Now, it is also becoming accessible to a broader population due to the declining price of solar panels and incentives from governments.

3.1.2 Historical Evolution of Solar Energy

Today's solar technology has developed through various stages in history. Studying this gives us an insight into how solar energy has molded present systems and how it ought to affect future energy and economic growth.

Ancient Uses: Passive Solar Design in Civilizations

Solar energy has been in use for light and heat for thousands of years. All ancient civilizations used solar energy for their daily needs. They cleverly use distinct designs to maximize sunlight exposure during winter and supply shading in summer to reduce dependence on external energy sources. These passive designs were very primitive, but they showed an understanding of the sun's potential and its importance in energy sustainability.

Photovoltaic Breakthroughs: Bell Labs (1954) to Modern Innovations

The solar revolution started with the development of photovoltaic technology by Bell Labs in 1954, which introduced practical silicon solar cells that were capable of converting sunlight directly into electricity. This breakthrough marked the beginning of a new era in energy production, paving the way for the widespread adoption of solar power. Its efficiency was just 6%, but its impact was much bigger, as it had opened the door for a new form of renewable energy without converting sunlight into heat to produce electricity. In the last few decades, research in material science has gained momentum, leading to improvements in PV (photovoltaic) technology and improved efficiency of solar cells by up to 20%. Recent developments, such as multi-junction cells (which use multiple layers that can capture a broader spectrum of sunlight) and thin-film solar cells, have opened new avenues of solar applications. Large-scale solar farms are producing hundreds of megawatts, while distributed solar systems (such as rooftop solar panels) have reduced the dependence on the central power grid by generating electricity for households and small businesses.

Key Milestones: Oil Crisis (1970s), Cost Reductions, and Policy Shifts

The oil crisis of 1970 was the turning point in exposing the vulnerabilities of relying on fossil fuels. This motivated the research and investment in the search for alternative energy sources. Governments across the globe were heavily motivated and gave incentives to promote the development of innovative technologies related to solar energy. Another important milestone was a reduction in the cost of energy. The rate of solar energy has become competitive with traditional energy. In many parts of the world, solar energy has become the cheapest energy. It is due to mass manufacturing, technological advancements, and support from governments and other organizations. Governments across the globe have realized the impact of climate change and have implemented many incentives, like tax credits and subsidies, to promote solar and renewable energy. The Kyoto Protocol and the Paris Agreement show humanity's intentions for the transition to renewable energy sources.

3.2 Fundamentals of Solar Energy

All our energies are ultimately received from the Sun. It requires a thorough understanding of the various terminologies and principles to harness solar energy. It is important to understand the interaction of solar radiation with the atmosphere to design efficient solar energy systems. Key concepts are solar radiation, photovoltaic effect, and concentrated solar power (CSP). Now, we shall discuss these concepts in detail in the following sections.

We have already studied in school that the sun emits energy in the form of electromagnetic radiation. Its wavelength ranges from 0.2 µm to 2.6 µm. It consists of visible light, ultraviolet (UV) radiation, and infrared (IR) radiation. However, the visible and infrared part bears the largest amount of energy in the full spectrum. Solar radiation at the Earth's surface is less intense than in the upper atmosphere because some wavelengths are scattered and absorbed by gas molecules like nitrogen, oxygen, water vapor, and carbon dioxide. The solar spectrum is essential in climate monitoring and weather forecasting.

- **Solar Irradiance**: Solar Irradiance may be defined as the power received per unit area from the Sun in the form of electromagnetic radiation. It is typically measured in watts per square meter (W/m^2).
- **Solar Constant**: Solar constant may be defined as the amount of solar irradiance received at the top of Earth's atmosphere. Its value is approximately 1,366 W/m^2. This

value is constant, but the amount of solar energy received at Earth's surface is reduced due to atmospheric conditions.

- **Spectral Distribution**: It is the energy distribution for different wavelengths of solar radiation. It is important in certain applications. For example, photovoltaic (PV) cells are designed to absorb certain wavelengths of light, while concentrated solar power (CSP) systems often focus on the entire spectrum. It is dependent on many factors, including atmospheric effects. Its energy density is reduced due to scattering, reflection, and absorption in the atmosphere. Aerosols and clouds also become key factors in reducing the solar radiation reaching the Earth's surface.

- **Insolation**: Insolation may be defined as the amount of solar energy received per unit area for a particular period. Thus, its value depends on weather conditions, the area's geographical location, and the time of year. It is measured in kilowatt-hours per square meter per day ($kWh/m^2/day$).

- **Peak Sun Hours**: It represents the equivalent number of hours per day having solar irradiance of 1,000 W/m^2. It means that a location having two peak sun hours is equivalent to receiving the same amount of energy as 2 hours of full sunlight.

- **Photovoltaic Effect:** It is the process by which sunlight is converted into electricity by using a semiconductor material. This effect is widely used in solar panels. The most used semiconductor material is silicon. The basic unit that generates electrical energy is known as the solar cell. The efficiency of solar cells depends on the design of the cell, the quality of the semiconductor material, and the material's bandgap (energy needed to free an electron from its orbit).

- **PN Junctions**: It is formed when we combine a p-type (positively charged) and an n-type (negatively charged) material. When sunlight falls on this junction, it excites the electrons, liberating electron-hole pairs, leading to an electrical current.

- **Electron Excitation**: It refers to the phenomenon of striking photons of sunlight to free electrons from their atomic bonds. These free electrons then create an electric current.
- **Electric Field**: The p-n junction formed in the silicon cell creates an electric field that separates electrons and holes so that they move in opposite directions. This leads to a potential difference across this junction through this electric field, which can be harnessed to power electric devices.

3.2.1 Photovoltaic (PV) Systems

A photovoltaic system is a widely used way to convert sunlight directly into electricity through solar panels. They are used in small-scale residential rooftop plants and large-scale solar farms. The following are the components of a PV system:

- **Solar Panels**: These are made up of multiple solar cells. Solar panels consist of numerous solar cells interconnected in a suitable arrangement to have the required voltage or current. When sunlight falls on these solar cells, it is converted directly into direct current (DC) electricity.
- **Inverters**: Our grid supplies alternating current (AC), which is used by most household appliances and industrial drives. These inverters are used to convert DC generated from solar panels to AC for either use in our appliances or to feed into the electricity grid.
- **Mounting Systems**: These are rigid structures to support solar panels so that they can safely stay intact for many years.
- **Batteries**: These are optional; they are needed only if local storage of generated electricity is required to be used during periods when no power is being generated due to the unavailability of sunlight, say, during the night. Now, most

installations do not use batteries and are directly feeding the excess generated electricity to the central grid. During the night or low sunlight, energy is accessed from the central grid.

- **Advantages**:
  - These are modular and scalable.
  - They need minimal maintenance.
  - They can be very easily installed in separate locations like rooftops, canal tops, arid lands, and remote areas.

- **Challenges**:
  - Its efficiency is affected by multiple factors, such as shading, dust accumulation, and temperature.
  - Its energy production is unreliable as it depends directly on weather conditions. So, it is highly intermittent and unreliable, particularly in the rainy and cloudy season.

> ***Activity***: *Open YouTube, see few videos to understand the generation of electricity using semiconductor material. Do you know, we can't see electricity, but we can feel its presence?*

3.2.2 Concentrated Solar Power (CSP)

This is a modified version of the traditional method to concentrate sunlight for heating purposes. As the name itself suggests, this method uses mirrors or lenses to focus sunlight from a wide area onto a small area. This concentrated sunlight in the form of heat is used to generate electricity. The concentrated solar power (CSP) method is used for very large-scale power plants. They also need regular maintenance.

Types of CSP Systems:

- **Parabolic Troughs**: This method uses curved mirrors to gather sunlight and focuses this sunlight onto a tube having fluid for heat transfer.
- **Solar Power Towers**: This method uses lots of mirrors (heliostats) to direct sunlight on the top of the central tower. This concentrated heat is used to heat a liquid to transfer heat for electricity generation.
- **Dish Engines**: This method uses parabolic dishes to concentrate solar energy at a focal point.

- **Advantages**:
  - The CSP system is more suitable for arid or non-fertile areas with strong sunlight. This barren land can be used for power generation.
  - Unlike the PV system, thermal energy can be stored, so electricity can be generated even in the absence of sunlight.
  - In comparison to PV systems and other traditional methods, CSP systems offer higher efficiency.
- **Challenges**:
  - CSP needs exceptionally large areas with strong direct sunlight. Hence, it is not economically viable for countries that have no barren land.
  - CSP requires a high initial investment and complex infrastructure.
  - Being spread in a large area with lots of components and many moving parts, it requires continuous maintenance.

3.2.3 Passive Solar Design

This is the most traditional way of harnessing solar energy. It focuses on designing buildings and materials for heating, cooling, or lighting. As the name itself suggests, this system does not use any mechanical system.

- **Orientation**: In areas with less sunlight, buildings are oriented to have maximum exposure to sunlight to get the largest amount of heat. In areas with strong sunlight, buildings are oriented to minimize sunlight to keep the houses cool.
- **Thermal mass**: Building materials like bricks or concrete absorb heat during the day and release it at night.

- **Insulation**: Heat loss can be reduced, or cooling can be maintained using insulation, thus improving energy efficiency.

- **Advantages**:
  - These are part of a system and require only proper design.
  - Since they are part of the design, no major additional cost is required.
  - These have low operating costs and minimal maintenance.
  - It is passive, thus reducing dependence on artificial heating or cooling and saving power.
  - They are environmentally sustainable.
- **Challenges**:
  - Passive methods require careful planning and design.
  - Its effectiveness depends on local climate and geographical location.

3.2.4 Comparative Analysis (Table: PV vs. CSP vs. Passive)

Features	**Photo Voltaic System (PV)**	**Concentrated Solar Power System (CSP)**	**Passive Solar Designs**
Energy Conversion	Direct sunlight to electricity	Sunlight to Heat and then Heat to Electricity	Sunlight to Heat or Light
Scalability	High	Large-scale only	Building specific only
Efficiency	15 to 22%	20 to 35 %	Design dependent
Storage	Batteries required	Thermal Energy Storage	Design dependent
Maintenance	Low	High	Very Low
Cost	Moderate	High	Low
Land	Moderate	High	Minimum

Table 3.1 Comparison of different methods for solar energy [18].

3.3 Photovoltaic (PV) Technology

As said in the earlier section, with the help of semiconductor material, we can easily convert solar energy into electrical energy.[22]. It has become popular as it is a clean, renewable, and sustainable source of energy. It is cheap, so it is now widely used in residential, commercial, and industrial applications, along with large-scale solar farms to generate electricity. Due to advancements in material technology, this design's efficiency has drastically increased from a meager 6% to 22% for Thin-Film (CdTe). However, Perovskite and tandem-type solar cells have reached a mark of more than 25 % and 33% in lab conditions. As the world is inclining toward sustainable energy sources, PV systems will play a pivotal role in leading the race with emerging technologies like perovskite cells, bifacial panels, and tandem cells. In this section, we shall focus on components of PV systems, their designs, and emerging technologies in this field [6]. Figure 3.1 represents the schematic internal diagram of a PV solar cell, while Figure 3.2 shows the typical layers of a solar panel.

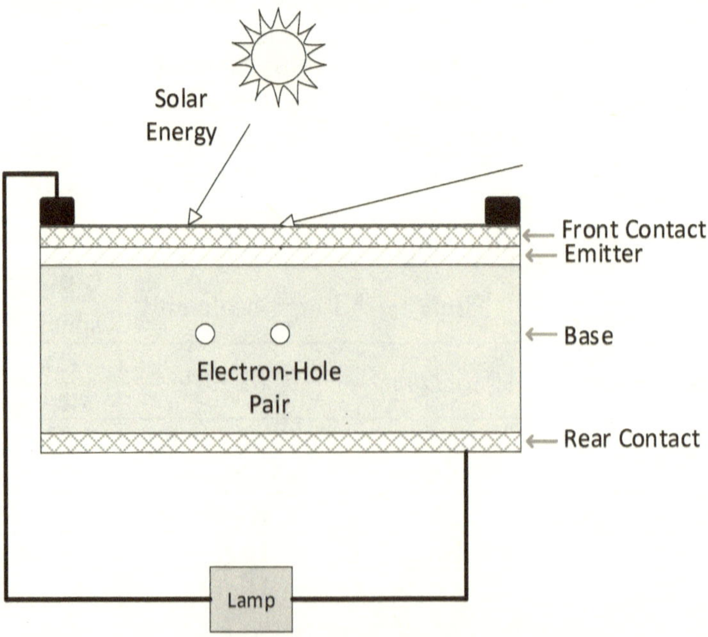

Figure 3.1 Schematic Internal Diagram of PV Solar Cell

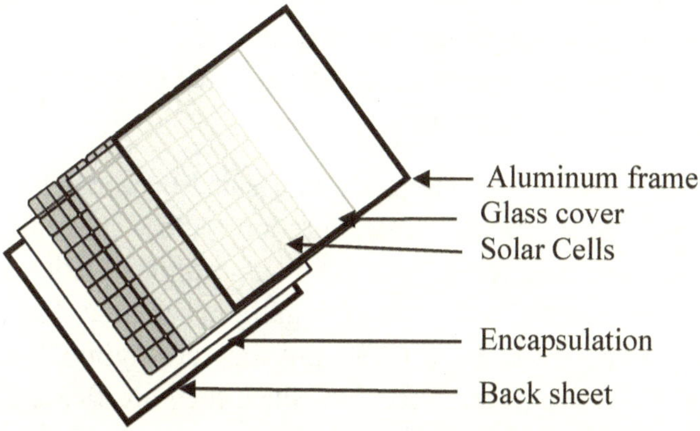

Figure 3.2 A schematic Diagram to show the Protection Layers of a PV Solar cell

3.3.1 Components of PV Systems

A typical Photovoltaic system consists of solar cells, modules, arrays, inverters, and a supporting or mounting system. These components work together to convert direct sunlight into DC electricity.

(a) Solar Cells:

These are the core components of PV systems, which are responsible for converting solar energy into electricity through the photovoltaic effect. Solar cells can be classified into three types:

1. **Monocrystalline Solar Cells**: You might have seen rooftop solar panels with a dark appearance in your city or college. Those are monocrystalline solar cells. They are made from a single crystal structure of silicon. They have become the popular choice for rooftop PV systems due to their high efficiency (15-20%) and longer life. They can perform well

even in low-light conditions, but are more expensive due to the complex manufacturing process.

2. **Polycrystalline Solar Cells**: These are made by melting multiple silicon crystals and are less efficient (13% -18 %) in comparison to monocrystalline, but are more cost-effective, so they are widely used in large-scale power plants where cost is a significant factor.

3. **Thin-Film Solar Cells**: These are fabricated by depositing layers of photovoltaic material on the substrate. Semiconductor materials may be amorphous silicon, cadmium telluride, or copper indium gallium selenide. Their efficiency is typically 10% -12%, but they are lightweight and less expensive to manufacture. Since they are flexible, they are preferred on curved surfaces.

(b) Modules, Arrays, and Balance of System (BOS)

1. **Modules**: The heart of the PV system is the Solar cell. They are sensitive and delicate and, hence, need protection to withstand harsh weather conditions. These solar cells are connected and encapsulated to form a solar module, commonly referred to as a solar panel. Modules protect the cells from environmental factors and supply structural support.

2. **Arrays**: An array is formed by connecting multiple modules. Its size depends on requirements. They can be installed on rooftops or the ground or can even be integrated with building walls. Figure 3.3 shows arrangements of solar cells, modules, and arrays in a panel.

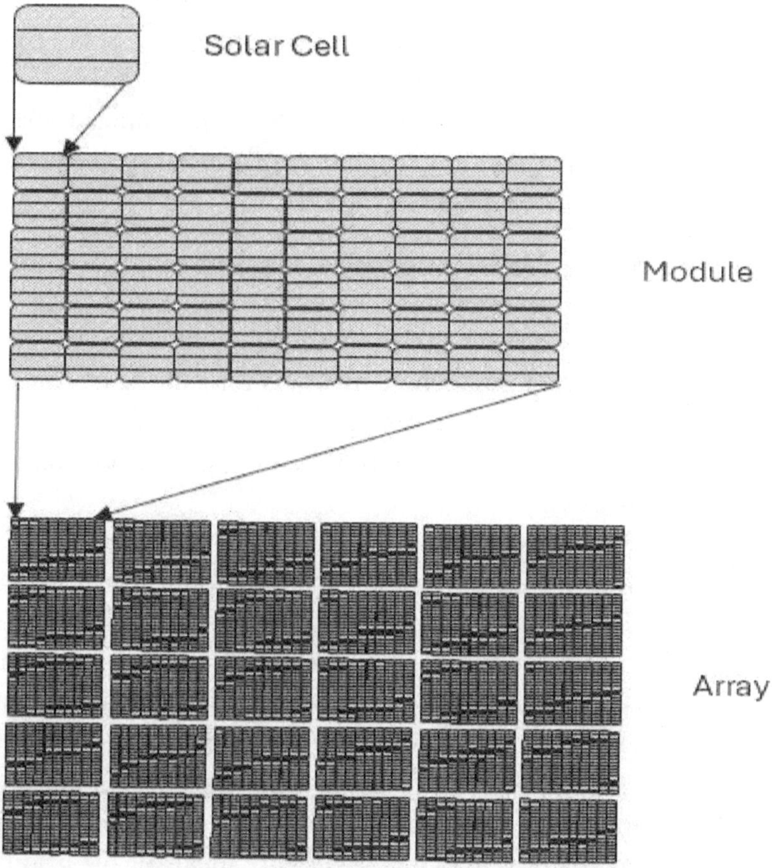

Figure 3.3 PV Solar Cell, Module, and Array in a Solar Power Plant

3. Balance of System (BOS): These are all items other than solar panels. It includes inverters, mounting systems, wiring, charge controllers (for off-grid systems), and energy storage systems (batteries). These components are the backbone of the system and ensure the efficient operation of the PV system along with integration with the electrical grid.

***Activity**: Visit a solar PV rooftop plan on your terrace. Touch the panels. Why are they so smooth? Observe the panel, try to identify the solar cell, array, and panel.*

(c) Inverters

Inverters convert the DC power generated by solar panels into alternating power to be fed into the grid. Recent developments in the field of power electronics have revolutionized the inverter industry, leading to the penetration of rooftop solar power plants into common households.

- **String Inverters**: In this case, a string of multiple solar panels is formed in series. The DC power generated from this string of panels is then converted into AC by a single inverter. Using single inverters for a string of panels is cost-effective but may become less efficient if a few panels are in shade or underperforming. This is the most common type of inverter.
- **Microinverters:** Contrary to string inverters, Microinverters are installed on each solar panel. So, it maximizes the power output, especially if some panels are underperforming, some panels are in shade, or some panels are in different orientations. This system maximizes efficiency and flexibility but is expensive.
- **Hybrid inverters:** These are systems with energy storage for later use. It combines the features of traditional inverters and battery inverters. These are ideal for off-grid or hybrid systems and are suitable for application in areas with an intermittent power supply.

(d) Mounting Systems

These are used to keep the solar panels and other systems in place, irrespective of wind and other conditions. They also optimize their exposure to sunlight. There are three main types of mounting systems:

- **Rooftop Mounting Systems**: They are mounted on the roofs of residential, commercial, or industrial buildings. Since the space of roofs is limited, these are designed to be space-efficient to reduce the need for space. They may be fixed or adjustable to optimize them to .tilt to maximize power production.

- **Ground-Mounted Systems**: These are installed on open ground or canal tops, where land availability is maximum. These are typically used in large-scale solar farms in deserts or fields. Since they are on the ground, their installations and maintenance are easy.

- **Tracking Systems**: These are used to increase the efficiency of solar farms. They use motors to adjust solar panels following the sun's position in the sky. Tracking systems use motors to adjust the position of solar panels throughout the day, following the sun's path. Single-axis trackers move panels along one axis (usually east-west), while dual-axis trackers adjust both the tilt and orientation. Tracking systems can increase energy production by up to 25% but are more expensive and require additional maintenance.

3.3.2 PV System Design and Optimization

To ensure long-term reliability and maximize energy production, it is necessary to optimize the design by considering system configuration, tilt, shading, orientation, supporting power electronics components, and efficiency.

- **Grid-Tied Systems:** Grid-tied systems are the most common systems as they are the most cost-effective option. These systems feed back excess power into the grid. Batteries are not provided in these systems; hence, they can not provide power during grid outages or blackouts.
- **Off-Grid Systems:** These systems are not connected to the electrical grid. They use batteries to store solar energy to be used during the night or in low sunlight. Since they require batteries, they are costlier, but they are ideal for remote locations.
- **Hybrid Systems:** These systems have batteries and are also connected to the grid. They store excess energy in batteries and use it back during the night or in low sunlight. Hybrid

systems are better suited for locations where one can find frequent power cuts.

- **Tilt and Orientation:** To have the maximum production of electrical power, it is necessary to align the solar panels towards the direct sunlight as close as possible to perpendicular to the direction of the incident sunlight. Since Earth is constantly rotating about its axis and revolving around the sun, the direction of sunlight keeps changing. During the day, panels are required to be moved from the East (in the morning) to the West (till evening) to follow the sun. Likewise, the direction of sunlight also varies in summer and winter. "Tilt" refers to the angle of the solar panel with respect to the ground, while "Orientation" refers to the direction of the solar panels with respect to true north. Tilt is the vertical angle, which affects how long the panel receives the sunlight at a given time, while orientation impacts how long the panel receives the sunlight throughout the day. The orientation of solar panels in the northern hemisphere should be true south, and in the southern hemisphere, it should be true north to receive the maximum direct sunlight.
- **Shading:** It is essential to conduct a proper site survey to ensure that obstructions like tall buildings, trees, and adjacent walls do not drop shadows on solar panels at the proposed site. Shading can reduce the efficiency of a PV system.
- **Efficiency:** De-dusting at regular intervals, proper wiring, regular maintenance, and high-efficiency panels can provide high efficiency for an extended period.

3.3.3 Emerging PV Technologies

Due to extensive research in the field of solar energy, new materials and designs are being developed to increase efficiency and reduce the cost. Newer applications are being assessed to expand the utility of solar panels, like solar-powered cars, solar-powered yachts, solar-powered planes, folding solar panels, transparent solar panels, etc. The following are the most promising emerging technologies:

- **Perovskite Solar Cells** are a special type of photovoltaic material that has high efficiency and low cost. They are made of organic ions, metal ions, and halogens, typically hybrid organic-inorganic lead (Pb) or tin (Sn) halide-based materials. The perovskite substrate is created by spinning the substrate in a pool of mixed precursor inks, and then the substrate is heated to crystallize the precursors and remove the extra solvent. These are easy, cheap, and simple to manufacture. They are flexible, semitransparent, and light in weight, and can have efficiency as high as 25%, so they are an attractive option for commercialization. However, certain challenges are yet to be resolved before its commercialization, such as stability, scalability, and environmental concerns related to lead content.
- **Bifacial Panels** can capture light on both sides of the solar panels. This leads to higher efficiency. These are more effective in regions of higher reflectivity, like sandy or snow-covered areas. Bifacial panels are costlier.
- **Organic Photovoltaic Cells** (OPV)offer advantages like being lightweight, low-cost, and flexible. They are made from carbon-based materials. Their efficiency is lower than traditional silicon-based cells, but it is expected to improve their efficiency and durability with research. OPV cells can be better suited where traditional silicon cells are cumbersome to use. OPV cells can be applied to surfaces such as windows, cars, or buildings to meet the local energy demand.
- **Quantum Dot (QD) Solar Cells** are nanoscale semiconductor materials that can absorb specific wavelengths of light. QDs in multiple layers can absorb different wavelengths of light, thus increasing the efficiency by absorbing a broader spectrum. QDs can generate multiple excitations from a single photon. They can retain their

efficiency even after long-term storage. They can be fabricated using many methods such as chemical baths, hydrothermal synthesis, and successive ionic layer adsorption and reaction (SILAR).

- **Tandem Cells** are made of two or more PV cells stacked together to use the sunlight spectrum more efficiently. To achieve this, they are made from different semiconductor materials having different band gaps. Their work involves basic but smart mechanisms. Top cells have large band gaps and convert UV and blue light into electricity, while the bottom cells have smaller band gaps and convert red and infrared (IR) light into electricity. To filter the light automatically, these cells are stacked in order of decreasing bandgap. They have higher efficiency and lower cost.

3.4 Concentrated Solar Power (CSP)

Concentrated Solar Power (CSP) is a renewable energy technology that involves focusing sunlight on a small area to produce heat. This heat is then used to generate electricity. We have seen in the previous section that a PV system directly converts sunlight into electricity. Unlike PV, CSP technology uses solar radiation to generate heat, and this generated heat is then used to drive a turbine to produce electricity. This indirect conversion has the advantage of thermal energy storage (TES), which helps us to produce electricity even when sunlight is not available, leading to improved grid stability. With advancements in technology and cost reductions, CSP technology is becoming a promising option to provide reliable renewable energy. Its application extends beyond electricity generation to industrial heating and desalination plants to provide drinking water in deserts.

> ***Activity**: Visit a solar PV rooftop plan on your terrace. Observe the reading of the solar meter to note today's reading. Visit the meter tomorrow at the same time, and note the meter reading. Calculate the difference. It will give you the solar generation of a single day. It should be approximately 4 to 5 units per kW per day. Suppose you have installed a 3-kW plant, then it should be around 13 to 15 units.*
> *Is it lower than that even on a sunny day? If yes, then it's time to clean the solar panels with water.*
>
> *1. Can you calculate the payback period of your solar plant? Assuming a maintenance-free life of 25 years,*
>
> *2. Calculate the total savings due to the solar plant (assume Rs 6 per unit).*

3.4.1 Principles of CSP Technology

The basic principle involved in the CSP technology to generate electricity is the thermal conversion of solar energy through a thermodynamic cycle. The main types of CSP systems are Parabolic Trough Systems, Solar Power Towers, Dish engines, and Linear Fresnel reflectors.

3.4.1.1 Parabolic Trough Systems

Parabolic Trough Systems are the most popular and commercially established CSP technology in the market. The main component of this system is a long, curved, parabolic-shaped mirror that focuses the sunlight on a long tube placed at the focal line of this curved mirror. Synthetic oil or any other heat transfer oil kept in this tube absorbs the heat, reaching temperatures up to 400°C. This thermal energy is then used to produce electrical energy through the steam turbine and generator. These are widely used in countries such as the United States, Spain, and India in utility-scale CSP plants. Parabolic trough systems have high efficiency, a proven history, and integration of thermal energy storage systems, but they require large land areas and regular maintenance of mirrors to have optimum efficiency. Figure 3.4 represents the basic working of a parabolic solar energy collector.

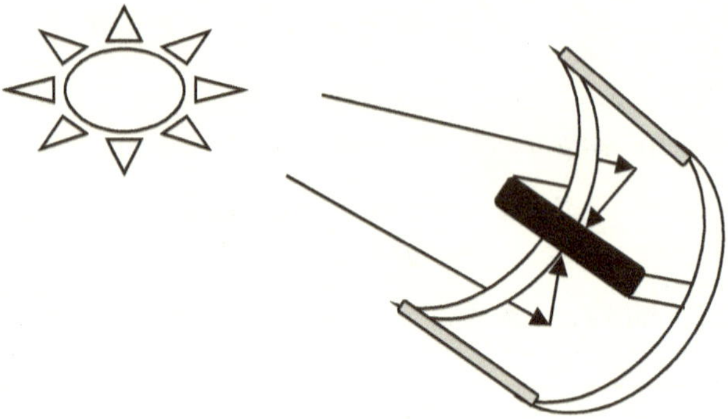

Figure 3.4 Parabolic System to Harness Solar Energy

> ***Activity**: Try to replicate the model of heat generation using a solar tower and mirrors. Can you feel the heat on the solar tower from the surrounding heliostats?*

3.4.1.2 Solar Power Towers

Solar power tower systems consist of a central tower, thousands of mirrors (known as heliostats) spread in a wide area of land, a working fluid, and a computer-controlled network to align the mirrors to track the sun from sunrise to sunset throughout the year. Solar radiation received on these mirrors is reflected to the top of a central tower. Thus, we may say that solar radiation received on a wide area is collected (through the reflection of these mirrors) on a small boiler located at the top of the central tower. This concentrated heat in the working fluid raises the temperature above 600°C. This heated fluid (often molten salt is then used to produce high-pressure steam to rotate a turbine to generate electricity. The main advantages of solar tower systems are their ability to achieve higher temperatures, higher efficiency, and thermal storage to generate electricity even during the night. The main demerits of these systems are a remarkably high initial investment, a wide land requirement, a complicated mirror tracking mechanism, and regular maintenance.

3.4.1.3 Dish Engines and Linear Fresnel Reflectors

These are also known as parabolic dish systems. They use the principle of the concave mirror to focus solar radiation at a focal point. The receiver placed at the focal point of the dish transfers the heat to a Stirling engine, which converts thermal energy to mechanical energy and then to electrical energy. Dish engines can operate independently and hence are more suitable for remote areas. The major challenges in its widespread use are excessive cost and regular maintenance.

Linear Fresnel Reflectors (LFRs) are a modified version of the parabolic trough system, but with lower cost and lower efficiency. This system employs slightly curved or flat mirrors arranged linearly to focus solar radiation on a stationary receiver. The LFR system is gaining popularity due to its straightforward design, lower cost, and the possibility of integrating with industrial applications.

3.4.2 Thermal Energy Storage in CSP

The main advantage of CSP technology is its ability to store thermal energy captured during the daytime and utilize it at times, even when sunlight is not available. This makes it possible to generate electricity at our convenience to increase grid stability and make renewable energy more dispatchable.

Molten Salt Storage and Phase-Change Materials

Molten salts have been widely used to store heat in various applications. Mostly, a mixture of sodium nitrate and potassium nitrate is used to store heat at temperatures ranging from 290°C to 600°C. This stored heat energy is then used to generate electricity during low sunlight or no sunlight, or during a period when electricity demand is higher.

Phase-changing materials, or PCMs, are another promising and emerging alternative option in the field of thermal energy storage.

These materials change their physical state from solid to liquid or liquid to gas and vice versa. They offer higher energy density with reduced losses. They may replace the traditional molten salt method if challenges about material degradation and higher costs are resolved for large-scale adoption.

3.4.3 Applications of CSP

CSP technology is becoming a prominent alternative in the field of renewable energy. CSP technology has been in use in a wide range of applications in utility-scale power generation and industrial heating.

- Utility-Scale Power Generation: Utility-scale power generation is one of the most significant applications of CSP. These plants are typically deployed in areas with high direct normal irradiance (DNI). Many countries, such as Spain and Morocco, have heavily invested in CSP technology-based projects to enhance their renewable energy mix. The main reason behind interest in CSP technology is its ability to store thermal energy, allowing it to function as a baseload power source. Its reliability can be further enhanced by hybridization with other sources like biomass and natural gas. At present, CSP has a higher initial cost than PV systems, but large-scale production may reduce the cost.
- Industrial Process Heat: Beyond electricity generation, CSP can also supply high-temperature heat for various industrial applications [5]. Many industries that require thermal energy for various chemical processing, metallurgy, and food production can avail themselves of heat energy exceeding 500°C from CSP.
- CSP can be useful in various other applications like desalination projects, evaporative processes, and enhanced oil recovery (EOR).

3.5 Integration, Storage, and Hybrid Systems

Solar energy is intermittent in nature. Its intensity increases from morning, reaches a peak in the afternoon, and reduces to zero after dusk till morning. The integration of such intermittent energy sources into the grid always presents complex challenges. As the global energy paradigm is shifting towards renewable energy sources in search of sustainability, addressing such issues becomes mandatory. In this section, we shall discuss grid integration challenges, energy storage, and solar hybrid systems.

3.5.1 Grid Integration Challenges:

To thrive on sustainable energy sources, the inclination towards renewable energy sources like solar energy is increasing. The increasing penetration of such intermittent sources always poses challenges to the stability and reliability of integrating them into the grid [13].

Intermittency and Duck Curve: The "duck curve" is a graphical representation of the gap between availability and the demand for energy throughout the day. It was first coined by the California Independent System Operators (CAISO) in 2012. It resembles the silhouette of a duck, which indicates that when solar generation is high, then net demand reduces, whereas when solar generation is low in the evening, demand increases rapidly. It becomes difficult to maintain the balance and grid stability during these hours. To address this challenge, an energy storage system (ESS) can be deployed. Large-scale batteries may be used to store excess energy during peak solar hours and feed back to the grid during peak demand hours. The Great Kiskadee Storage project in Texas is a 100 MW battery storage project and an ideal example to demonstrate grid stability using ESS. Another way to mitigate this challenge is to charge differently for off-peak and peak demand hours. Incentives to use energy during off-peak hours will reduce the demand for peak hours, thus stabilizing the grid. Grid modernization is also necessary to increase flexibility and manage fluctuations in energy supply and energy demand.

Smart Grids:

Smart grids incorporate the latest digital technology, and they keep updating with technology. They integrate the traditional electrical system with digital technology. They enable prompt response and two-way communication between utilities and consumers with real-time data monitoring. The main components of the smart grid are:

- Advanced Metering Infrastructure (AMI): Smart meters provide real-time data about energy consumption, letting consumers make decisions and utilities optimize the grid.
- Automated Control Systems: They can immediately respond to grid disturbances, help in rerouting the power, and prevent power outages.
- Integration of Distributed Energy Sources (DERs): Various sources like rooftop solar power plants, wind turbines, and others can be seamlessly integrated with smart grids, enhancing grid resilience.

3.5.2 Energy Storage Solutions

Effective energy storage is pivotal for managing the intermittency of renewable energy sources. Advancements in storage technologies are enhancing the capacity to store and dispatch energy as needed.

Lithium-Ion Batteries:
Lithium-ion batteries are dominating the market of energy storage systems due to their higher energy density and lower weight. Its cost is comparatively higher, but recent developments have significantly reduced its cost, making it possible to use it in off-grid and on-grid applications. Owing to its lightweight, it is widely used in electric vehicles.

Flow batteries:
They use liquid electrolytes in external tanks, which offer longer life and scalable energy storage solutions. This makes them suitable for applications of long-duration storage.

Pumped Hydro Storage (PHS):
A pumped hydro storage system is a traditional way of meeting the peak load by utilizing the excess energy of a low load. Water is pumped back to the reservoir at higher altitudes during low load or when there is excess energy available. This elevated reservoir water

is then used to generate electricity to meet the demand of the peak load. This system has geographical constraints and can not be used universally for all locations.

Thermal Energy Storage (TES):
In this system, heat energy is stored in certain materials during the availability of excess energy. This excess energy stored in heat form is then converted to electricity or may be used for heating purposes when the need arises. Its efficiency depends upon the material used for heat storage and the leakage of associated heat. However, innovations in materials and designs are improving the efficiency of TES. This system has been widely used in solar thermal power plants. *[Recall the solar tower section discussed in earlier sections of this chapter.]*

3.5.3 Hybrid Energy Systems:

Hybrid energy systems are combinations of multiple energy systems. They are designed to increase the reliability and efficiency of the overall system. Solar energy is not available during the night and the rainy season. To fulfill this energy gap and increase the reliability of the energy system, it is clubbed with a diesel or wind energy system. This topic has been discussed in detail in the chapter on wind energy. Brief information is provided here.

- Solar-Wind Hybrid system: Solar and wind energy complement each other. During the day, we have sunlight but low wind speed. During the nights, we do have strong, consistent winds but no sunlight. Thus, solar generation is more reliable during the day, while wind generation is more reliable during the night. Taking advantage of this, a combo system or, say, a hybrid system of solar and wind will be more balanced and more reliable to increase the system's stability.

- Solar-diesel hybrid system: This system is more suitable in regions of limited grid connectivity. During the day, excess solar energy can be converted into electricity, and during the night, diesel generators can be used to generate electricity.,

Thus, a minimum quantity of diesel is used, reducing greenhouse gas emissions. It is more financially affordable than batteries. They may be used to store excess electrical energy generated during the daytime.

3.5.4 Microgrids

A microgrid is a small-scale electricity network that connects consumers and the electric supply. It is a distributed energy resource that can operate independently or with the main electrical grid. They are more suited for remote or rural areas where the conventional grid is impractical or economically unviable. They incorporate mostly renewable energy sources, storage systems, and smart network management to have reliable energy availability. Microgrids offer many advantages. They function as stopgap arrangements for remote and rural areas to enhance healthcare, education, and other opportunities in off-grid regions. In most cases, they are autonomous in operation, so they offer great reliability in case of power outages during grid failures. Since they mostly use renewable energy sources, they help reduce the dependency on fossil fuels and contribute to decarbonization.

Microgrids have been equipped with Artificial Intelligence and IoT tools to enhance grid efficiency and optimize energy distribution. Recent developments in the field of batteries and hydrogen storage have increased the performance of microgrids.

3.6 Economics, Policy, and Market Trends

In this section, we shall discuss cost analysis, LCOE, financial incentives, policies, and a few international agreements. We shall also discuss solar growth and innovation from the perspective of China's solar dominance.

3.6.1 Cost Analysis

For any system to survive in the market, cost competitiveness is a crucial factor. Over the last few decades, the price per kW of solar energy has been significantly declining due to its widespread acceptability and production, cheaper innovative methods of manufacturing, and government support through a strong policy framework. The graph below (Figure 3.5) reflects the declining trend in solar energy prices. At present, in 2025, the capital cost per kW of solar energy for an on-grid solar system is approximately Rs 72,000, and for the off-grid solar system, it is Rs 80,000. Several components can lead to minor variations due to numerous factors. For example, higher quality components may have higher efficiency and life span but may be costlier. Cost may also be affected by required infrastructure, mounting, roof type, government incentives, subsidies, tax concessions, etc.

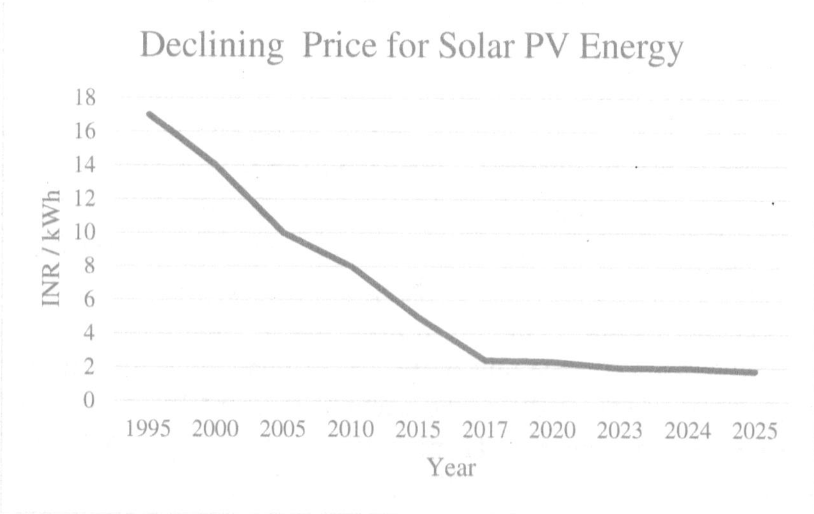

Figure. 3.5 Declining trend of Solar energy in the last three decades [50]

Levelized Cost of Electricity (LCOE) is a useful tool to judge the economic viability of any energy source. It gives us an idea about the per-unit cost of electricity generation, considering the capital costs, operational costs, maintenance, fuel costs (zero in solar PV), financial burden, and upgradation of parts (if required) during the maximum possible lifespan of the installed system. With the help of LCOE, we can compare the per-unit cost of different energy sources. It can be calculated using the following formula:

LCOE is calculated using the formula:

$$LCOE = \frac{\sum(C_t + O_t + F_t)(1+r)^{-t}}{\sum E_t(1+r)^{-t}}$$

Here,

C_t =Capital expenditures in year t

O_t = Operation and maintenance cost in year t

Ft = Fuel costs in years t (zero for solar PV)

E_t = Electricity generated in years t

r = Discount rate

t = Number of years

Solar energy does not require any additional fuel for its operation, it requires minimal maintenance, and has a considerably long life span of 25 – 30 years. Advanced manufacturing technology has been able to manufacture solar cells with higher efficiency and lower costs. Thus, solar energy has become cheaper than traditional fossil fuel-based energy sources. The following graph (Figure 3.6) shows the cost comparison for one unit of electricity generation for different energy sources. As of 2025, the per unit price of PV solar-based electricity has touched its all-time low of Rs 2.5 per unit. It is estimated that by the end of 2030, it may drop below Rs 1.9 per kWh. It's a booming sector. Won't you jump in to grab the opportunities to become a businessperson?

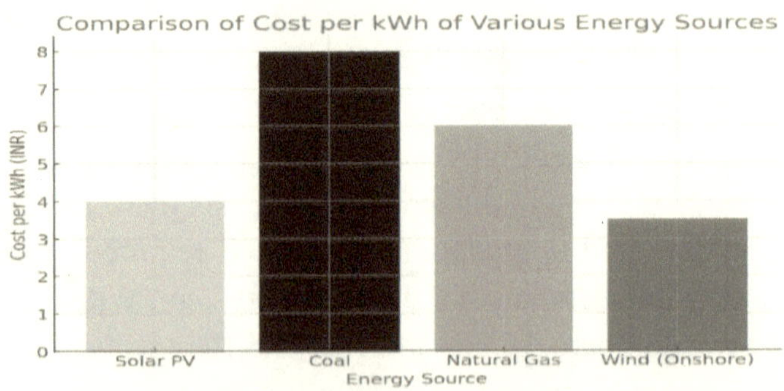

Fig. 3.6 Cost Comparison for Different Energy Sources

According to the International Renewable Energy Agency (IRENA), the global average LCOE for solar electricity has reduced to Rs 4.0 per kWh (in 2021) from Rs 31.50 per kWh (in 2010). This is a reduction of more than 87% in just 20 years. The following table shows the LCOE price comparison of various energy sources specific to India. LCOE prices per unit in India are comparatively much lower due to lower labour costs, strong government support, and an easy land acquisition policy. The global average LCOE for solar energy per unit is Rs 5.48 per kWh as of 2024.

Sr No.	Energy Source	LCOE Price
1	PV Solar Cells	Rs 2.5 per kWh
2	Onshore Wind	Rs 2.5- Rs 5.0 per kWh
3	Offshore Wind	Rs 7.47 per kWh
4	Natural gas	Rs 3.50 – Rs 8.0 per kWh
5	Coal	Rs 4.0 – 12.0 per kWh

Table 3.2 Energy Sources and their LCOE Price.

Major factors affecting the declining cost of Solar PV are:

- Manufacturing and Technological advancements: Due to innovations in metallurgy, improved designs, and advancements in semiconductor technology, the production of solar cells has resulted in higher efficiency, longer life, and lower costs.
- Mass production: Wide-scale acceptability and impacts of climate change have motivated people and governments to adopt clean energy, leading to a great demand for PV solar cells. This has led to mass production and hence a reduction in the price of solar panels.
- Development in power electronics and integrated chip technology has reduced the size of inverters, reduced their cost, and enhanced efficiency.
- The availability of easy financial mechanisms and lower interest rates made solar energy more accessible.
- Seeing the ill effects of climate change and global warming, countries across the globe have understood the importance of clean and green energy. So, they are motivating people

and companies to install large solar farms, and domestic rooftop solar systems through various schemes like PM-Kusum *(Pradhan Mantri Kisan Urja Suraksha evam Utthan Mahabhiyan)*, tax incentives, subsidies, and policy reforms have created a demand for solar panels on a mass level, leading to a drop in price.

3.6.2 Impact of Positive Policies and Financial Incentives

As said in the last section, seeing the negative impacts of climate change and global warming, governments across the globe have started to promote clean and green energy to have sustainable development through various financial and policy incentives.

- **Tax Credits:** Many countries, like India and U.S. offer relaxations in tax to individuals and companies to motivate people to adopt solar energy. This reduces tax liability and increases profitability.
- **Feed-in Tariffs (FiTs)** guarantee fixed revenue per unit of solar energy fed into the grid.
- **Net Metering** policy allows consumers with rooftop solar installations to export surplus solar energy to the grid, generating revenue for them. This concept has been widely appreciated and accepted by the people of India, enhancing rooftop solar installations.
- **Power Purchase Agreement (PPA)** ensures a fixed rate of solar energy fed into the grid by consumers. It may be with an individual or with a solar farm company.
- **Renewable Portfolio Standards (RPS)** require utilities to either generate or purchase a part of the electricity needed through renewable energy sources. Distribution companies (DISCOMs) track the record of energy procurement, say, from solar energy.
- **International Organizations** like the Paris Agreement play a significant role in the deployment of renewable energy sources. For example, the International Solar Alliance (ISA) was founded with the intention of facilitating finance,

technology exchange, and policy coordination among the member countries to promote solar energy [16].

3.6.3 Major Solar Farms across the World

Sr No.	Name	Capacity in MW	Technology	State/ Province	Country	Year of Est.
1	Bhadla Solar Park	2245	Photovoltaic (PV)	Rajasthan	India	2017-2020
2	Huanghe Hydropower Hainan Solar Park	2200	Photovoltaic (PV)	Qinghai	China	2020
3	Pavagada Solar Park	2050	Photovoltaic (PV)	Karnataka	India	2019
4	Mohammed bin Rashid Al Maktoum Solar Park	2000 + 5000 planned	Photovoltaic (PV) + Concentrated Solar Power (CSP)	Dubai	UAE	2013-in progress
5	Benban Solar Park	1650	Photovoltaic (PV)	Aswan	Egypt	2019
6	Tengger Desert Solar Park	1547	Photovoltaic (PV)	Ningxia	China	2017
7	Noor Abu Dhabi	1177	Photovoltaic (PV)	Abu Dhabi	UAE	2019
8	Kurnool Ultra Mega Solar Park	1000	Photovoltaic (PV)	Andhra Pradesh	India	2017
9	Villanueva Solar Park	828	Photovoltaic (PV)	Coahuila	Mexico	2018
10	Copper Mountain Solar Facility	802	Photovoltaic (PV)	Nevada	USA	2010

Table 3.3 Major Solar Farms in the World.

3.6.4 Major Solar Farms across India

Sr No.	Name	Capacity in MW	Technology	State	Year of Esta.
1	Bhadla Solar Park	2245	Photovoltaic	Rajasthan	2017-2020
2	Pavagada Solar Park	2050	Photovoltaic	Karnataka	2019
3	Kurnool Ultra Mega Solar Park	1000	Photovoltaic	Andhra Pradesh	2017
4	Rewa Ultra Mega Solar Park	750	Photovoltaic	Madhya Pradesh	2020
5	N P Kunta Solar Park	978	Photovoltaic	Andhra Pradesh	2016- in progress
6	Mandsaur Solar Farm	648	Photovoltaic	Madhya Pradesh	2017
7	Adani Kamuthi Solar Project	648	Photovoltaic	Tamilnadu	2016
8	Kamalanga Solar Park	500	Photovoltaic	Odisha	2022
9	Charanka Solar Park	500	Photovoltaic	Gujarat	2012
10	Dholera Solar Park	700 + 5000 planned	Photovoltaic	Gujarat	2021- in progress

Table 3.4 Major Solar Farms in India.

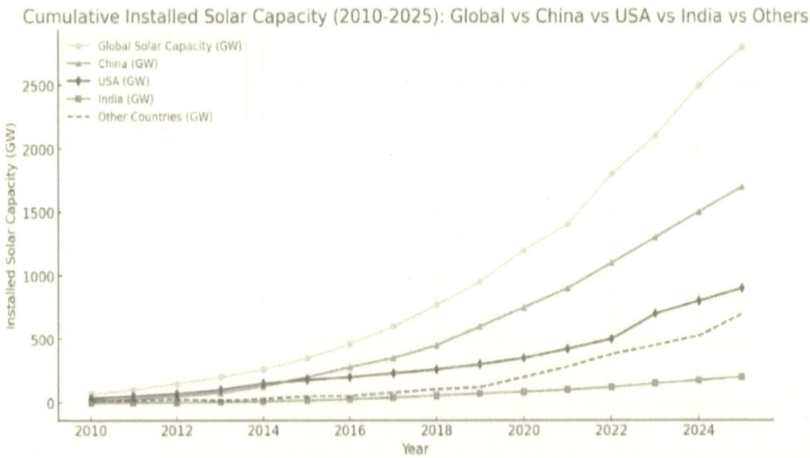

Fig. 3.7 Growth of Installed Solar Power Capacity in the last 15 years.

Dholera Solar Park, Gujarat, India, is set to become the world's largest solar park with a capacity of more than 5500 MW. As of February 2025, India has already surpassed 100 GW of solar installed capacity out of 218 GW of non-fossil fuel energy, thus solar energy is contributing almost half of the renewable energy. India is expected to contribute 292 GW of solar capacity to the total 500 GW of its renewable energy target. At the international level, China and India are leading the solar energy race. Global solar installation surpassed 2800 GW recently, with an annual growth of 500 GW per year. China has 887 GW installed solar capacity, which is 45% of global solar capacity [4]. Figure 3.7 tells the ever-expanding story of solar power.

3.7 Environmental and Social Impacts of Solar Energy

- **Environmental Benefits:** Like other renewable energy sources, solar energy also has the advantage of reduced greenhouse gas emissions. It does not require any fossil fuel for energy generation and does not involve any rotational mechanism like generators; hence, it does not produce any greenhouse gases during electricity generation throughout its lifetime. Unlike wind energy, it neither produces noise nor poses any threat to birds and bats. Unlike thermal and nuclear power plants, solar power plants (PV) do not require water for their operation; thus, they may be deployed in water-scarce and arid regions like deserts. So, in summary, we may say that solar power plants are environmentally friendly and may prove helpful in mitigating the ill impacts of climate change.
- **Social Impacts: The** Solar energy sector is providing opportunities for jobs for the skilled, semi-skilled, and unskilled workforce. Employment areas range from research, manufacturing, operation, maintenance, and management. Thus, it may improve the living standard of rural and urban areas. Solar farms are deployed in rural and remote areas, alleviating the energy deficiency of that area,

thus uplifting the living standard, and opening the doors of development through electricity access. Solar farms may generate revenue for farm owners. Rooftop solar power plants also utilize the idle space of roofs to generate electricity and revenue for rural as well as urban populations. It also provides resilience against power outages, thus increasing energy reliability.

- **Recycling:** Since energy generation through solar farms and rooftops is exponentially rising, it has raised concerns about the waste disposal and recycling of solar panels. Solar panels have a useful lifespan of about 25-30 years. An effective recycling mechanism must be mandated to recycle the solar panels to recover Silicon. Silver and Aluminum. An Extended Producer Responsibility (EPR) policy should be implemented to mandate that manufacturers manage the solar panels and inverters after the end of their useful life. Research in this field is underway to increase the useful life of panels, to prepare thin solar panels to reduce material use, and foldable and customized panels to suit various applications like car roofs and building walls.

3.8 Future Trends in Solar Energy

- Over the last decade, a new concept of **Building Integrated Photovoltaics (BIPV)** has emerged. Unlike traditional solar rooftop panels, BIPV serves a dual purpose to generate electricity as well as to function as a construction material. These are integrated with building structures such as roofs, facades, and windows. Technological innovations have made it possible to have lightweight, transparent, and aesthetically appealing panels to inspire people to adopt these solar panels.
- To reduce land use and maximize the use of solar energy, floating solar panels are installed on water reservoirs and lakes. Many times, canals are also covered with solar panels to serve the dual purpose of electricity generation as well as the reduction of water evaporation. In the case of

Agrivoltaics, crops are cultivated beneath the solar panels to maximize land use, reduce water evaporation, and mitigate the heat stress to enhance crop production.

- Countries like Japan, USA, and China are exploring the concept of **Space-Based Solar Power (SBSP)**, which involves the transmission of captured solar energy through satellites orbiting the Earth. This technology has the benefit of uninterrupted power irrespective of weather and day and night cycles. This technology is still in the experimental stage.
- Solar energy is also being utilized in electrolysis. Electrolysis is the process of splitting water molecules into hydrogen and oxygen through electricity generated using solar panels. Since this process does not use any fossil fuel energy source, the generated hydrogen gas is known as Green Hydrogen, which is emerging as a major player in the field of energy storage, the transportation sector, and industries.
- **International Solar Alliance (ISA):** It was launched for the promotion of solar energy worldwide. Presently, this organization has more than 120 member countries. The major aim of this alliance is to improve solar efficiency, to harness maximum solar energy, to reduce the per-unit cost of solar energy, and to increase energy access for remote and rural areas. Do you know, it was jointly launched by India and France in 2015?

3.9 Case studies

- **Noor Ouarzazate CSP Complex**, located in Morocco, is the world's largest concentrated solar power project with an installed capacity of 580 MW [6]. It is an ideal example for the better utilization of arid/desert land. It uses parabolic troughs and solar towers to generate electricity. Since it is based on CSP technology, it also provides the facility of thermal storage and nighttime electricity generation. This project has significantly contributed to meeting the renewable energy goals of Morocco.

- **Tesla Solar Roofs and Powerwall** are an innovative approach that aims to fulfill the energy needs of the residential sector. In Tesla solar roofs, PV cells are integrated with roof sheets, which are durable and aesthetically pleasing. Powerwall is used to store energy generated during the day to be used at night. This concept helps the homeowner to generate, store, and use solar energy very efficiently, reducing the dependence on the central grid.
- **India's Mega Solar Parks:** Through the large-scale power projects like Bhadla Solar Park, having 2250 MW, and Pavagada Solar Park, having a capacity of 2000 MW, India has shown the potential of solar energy. Positive policy, motivating governance, and easy finance can boost the solar energy sector, leading to accelerated growth in employment opportunities, infrastructure development, the manufacturing sector, and diminishing energy poverty, ensuring sustainable growth.
- **Solar Impulse Project:** This project was initiated by Swiss explorers Bertrand Piccard and André Borschberg. For the first time in human history, they demonstrated a global flight with the help of the Solar Impulse 2 aircraft. This aircraft was equipped with more than 17,000 PV solar cells. Energy generated by these solar cells was used to fly the aircraft. It paved a path for innovators to explore the options of renewable-powered air mobility.

Activities:

1. *Try to create a small charger for your own mobile using small flexible solar panels.*
2. *Try to build a simple solar powered setup to purify dirty water using evaporation and condensation. Use mirrors to boil the water.*

-:EXERCISE:-

Section A: Short Answer Questions

1. Define Solar Constant. State its approximate value.
2. Define 'Photovoltaic effect.'
3. List materials most commonly used in PV cells.
4. List the benefits of PV systems.
5. Define Peak Sun Hours.
6. List any two challenges faced by CSP systems.
7. Recall the typical efficiency of monocrystalline solar cells.
8. List materials used in the Thin-Film solar cells.
9. Draw PV cell, Array, Module, and Solar panel. Define each term.
10. Define LCOE in the context of solar energy.
11. State financial policies to promote solar energy adoption by households and small organizations.
12. List the top five countries leading the world in installed solar capacity.
13. List the top five states of India in installed solar capacity.

Section B: Long Answer Questions

1. Compare photovoltaic systems and Concentrated Solar Power systems based on energy conversion method, scalability, efficiency, maintenance, and cost.
2. Explain the working of a photovoltaic system. Also, state the role of its major components, like the inverter and batteries.
3. Discuss passive solar design. Appreciate its role in the heating or cooling of a building.
4. Explain various solar cells available in the global market.
5. Discuss how solar radiation interacts with Earth's atmosphere.
6. Explain the working of PV solar cells.
7. List limitations of rooftop solar power plants.
8. Describe the challenges of renewable energy with the grid.
9. Discuss the role of solar energy in rural development.

10. Explain the role of the Government of India in promoting rooftop solar power plants.
11. Differentiate between Grid-Tied and Off-Grid PV systems with suitable examples.
12. List the factors for the declining cost of solar energy.
13. State the importance of a Solar tower to provide non-intermittent energy.
14. Discuss the role of ISA in promoting solar energy.
15. Explore the available online and offline resources to compare the cost per unit for various energy sources in India.

Section C: Multiple Choice Q and A

1. Solar constant value is ~...............
   A) 66 W/m² B) 1,366 W/m²
   C) 366 W/m² D) 2,360 W/m²
2. .... part of the solar spectrum carries the maximum energy?
   A) X-rays
   B) Visible and infrared
   C) Radio waves
   D) Gamma rays.
3. A photovoltaic effect is
   A) Conversion of heat to electricity
   B) Conversion of sound energy
   C) Conversion of sunlight to electricity using semiconductors
   D) Reflection of solar radiation.
4. A most commonly used semiconductor in solar cells is
   A) Gold B) Silicon C) Ozone D) Gallium
5. A PN junction in a solar cell....
   A) Increases temperature
   B) Filters infrared radiation
   C) Separates electrons and holes to generate current
   D) Stores electricity
6. The "peak sun hours" refer to:
   A) Hours with sunlight above 3000W/m²
   B) Hours with sunlight above 2,000 W/m²C
   C) Equivalent number of hours per day with 1,000 W/m²

solar irradiance
D) Total sunlight received in a year.

7. A........... PV cell type is made from a single silicon crystal.
A) Silicon T B) Monocrystalline
C) Silicon Y D) Multicrystalline
8. In .... technology, heliostats reflect sunlight on a central tower.
A) Parabolic trough B) Flat plate collector
C) Solar power tower D) Dish engine
9. In which of the following solar technologies is thermal energy stored to generate electricity even without sunlight?
A) CSP B) PV
C) Passive solar D) Solar cookers.
10. The monocrystalline solar cells have .... efficiency.
A) 1–5% B) 5–10%
C) 10–15% D) 15–22%
11. The inverter most suitable for each solar panel individually is:
A) Hybrid inverter B) Central inverter
C) Microinverter D) String inverter.
12. Balance of System (BOS) does NOT include:
A) Batteries B) Solar cells
C) Mounting systems D) Inverters
13. The “insolation” is measured in?
A) Watts per hour
B) Watts per hour meter
C) Kilowatt-hours per square meter per day
D) Joules per square
14. The flexible thin-film solar cells are made from
A) Lithium-ion B) Cadmium telluride
C) Lead D) Iron oxide
15. The mounting system to adjust solar panels according to the sun’s position is
A) Rooftop B) Ground-mounted
C) Tilt-fixed D) Tracking
16. The challenge/s of photovoltaic systems are:
A) Dust accumulation B) Temperature effects
C) Shading D) All of the above

17. The main purpose of insulation in passive solar design is
    A) Increase sunlight exposure
    B) Store thermal energy
    C) Reflect solar radiation
    D) Reduce heat loss or maintain cooling
18. An advantage of CSP systems over PV is:
    A) Thermal energy storage B) No maintenance
    C) Lower initial cost D) Higher land efficiency
19. The lab efficiency of tandem-type solar cells is:
    A) < 10% B) < 22% C) >25% D)> 33%
20. A key advantage of passive solar design is:
    A) Requires batteries
    B) Uses mirrors and lenses
    C) Environmentally sustainable
    D) High operating cost
21. The basic principle of Concentrated Solar Power (CSP) technology is:
    A) Photovoltaic conversion of solar energy
    B) Thermal conversion of solar energy through a thermodynamic cycle
    C) Electrical conversion of solar energy
    D) Chemical conversion of solar energy
22. Long curved parabolic mirrors are used to focus sunlight on a tube containing heat transfer oil. Specify the type of CSP system:
    A) Dish Engine B) Parabolic Trough System
    C) Ray Tube D) Solar reflector
23. In Parabolic Trough Systems maximum temperature of oil may be:
    A) 250°C B) 350°C C) 400°C D) 550°C
24. The....... tracks and reflects sunlight to the central tower.
    A) Stirling mirrors B) Fresnel lenses
    C) Heliostats D) Geliostats
25. The .... is used in Solar Power Tower systems to transfer heat at high temperatures.
    A) Water B) Liquid Ammonia
    C) Molten salt D) Synthetic oil

26. A min merit of Solar Power Tower systems over Parabolic Trough Systems is
    A) Lower initial investment
    B) Higher temperatures and higher efficiency
    C) Smaller land requirement
    D) Simpler mirror tracking system
27. Dish Engines convert solar thermal energy into electrical energy via:
    A) Gas turbine engine B) Diesel engine
    C) Stirling engine D) Steam engine
28. Linear Fresnel Reflectors differ from Parabolic Trough Systems because they:
    A) Use a single large parabolic mirror
    B) Use multiple flat or slightly curved mirrors arranged linearly
    C) Cannot be integrated with industrial applications
    D) Use Stirling engines
29. A main advantage of solar tower CSP is
    A) Almost zero land
    B) Almost zero water use
    C) Ability to provide power even at night
    D) Ability to shut off in case of emergency.
30. A solar tower type CSP uses.... working fluid to transfer the heat.
    A) Calcium carbonate and magnesium oxide
    B) Sodium carbonate and potassium oxide
    C) Sodium chloride and potassium chloride
    D) Lithium chloride and ion chloride.
31. The cost per kW for an on-grid solar system is:
    A) Rs 50,000 B) Rs 60,000 C) Rs 70,000 D) Rs 80,000
32. Levelized Cost of Electricity measures:
    A) Capital cost only
    B) Maintenance cost only
    C) Per unit cost of electricity generation, considering all expenses
    D) Per unit cost of all expenses except transportation of fuel
33. Which cost component is zero for solar PV in the LCOE formula?

A) Operation cost B) Capital expenditure
C) Salary cost D) Panel cleaning cost

34. The per-unit price of PV solar-generated electricity is:
A) Rs 1.5 B) Rs 2.5 C) Rs 3.5 D) Rs 4.5
35. The approximate price per unit for solar in 2030 may be:
A) Rs 2.5 B) Rs 2.0 C) Rs 1.5 D) Rs 1.0
36. The global average LCOE for solar electricity from 2010 to 2021 is:
A) 55% B) 65% C) 77% D) 87%
37. The declining cost of solar PV is not associated with:
A) Mass production B) Increased fuel costs
C) Tax concessions D) Government subsidies
38. Feed-in Tariffs may assure:
A) Subsidized installation costs
B) Fixed revenue per unit of solar energy fed into the grid
C) Tax credits
D) Free solar panels to consumers
39. .... had jointly launched the International Solar Alliance (ISA):
A) India and USA B) India and France
C) China and India D) USA and France
40. The .... is the largest solar park in India.
A) Dessert Solar Park
B) Kurnool Ultra Mega Solar Park
C) Bhadla Solar Park
D) Tamil Giga Solar Park
41. .... has the highest installed solar capacity globally:
A) UAE B) India C) China D) France
42. Solar power plants do NOT require:
A) Fossil fuels B) Water for operation
C) Moving mechanism D) All of the above
43. The social benefit associated with solar energy is:
A) High water consumption B) Job creation
C) Increase in fossil fuel use D) Noise pollution
44. The life of solar panels in years is:
A) 10-15 B) 15-20 C) 25-30 D) 40-60
45. The uniqueness of Building Integrated Photovoltaics is:
A) Only generate electricity
B) Act as construction materials and generate electricity

C) They are very cheap
D) Are only installed on the ground.

46. Floating solar panels are installed to:
A) Increase land use B) Reduce water evaporation
C) Both A and B D) To save fish from water.

47. Space-Based Solar Power (SBSP) offers:
A) Intermittent power supply
B) Power transmission via satellites for continuous supply
C) Safe journey to the moon.
D) Communication

48. Green hydrogen is produced by:
A) Wood Combustion
B) Electrolysis using solar energy
C) Nuclear fission reaction
D) Biofuels

49. The Noor Ouarzazate CSP system uses:
A) Photovoltaic cells
B) Parabolic troughs and solar towers (CSP)
C) Nuclear power
D) Wind turbines

50. Tesla Solar Roofs combine solar PV cells with:
A) Roof shingles B) Wind turbines
C) Fossil fuel backup systems D) Coal generators

Answer Keys:

1 B, 2. B, 3. C, 4. B, 5. C, 6. C, 7. B, 8. C, 9. B, 10. D, 11. C, 12. B, 13. C, 14. B, 15. D, 16. D, 17. D, 18. A, 19. D, 20. C, 21 B, 22. B, 23. C, 24. C, 25. C, 26. B, 27. C, 28. B, 29. C, 30. B,31.B, 32.C, 33.D, 34.B, 35.B, 36.D, 37.B, 38.B, 39.B, 40*.C, 41.C, 42.A, 43.B, 44.C, 45.B, 46.C, 47.B, 48.B, 49.B, 50.A

*Bhadla Solar park 2245 MW capacity

वायो याहि प्रशस्तया धेनुर्हितं मिमिक्षती ।

वृषण्वतस्तुजे वसुम् ॥

Rigveda (1.134.4)

O Vayu (Wind), come with your blessings. Like a cow giving milk, let you bring prosperity and energy.

Chapter 4

WIND ENERGY

Learning Outcomes:

After reading this chapter, the learner will be able to:

1. Trace the evolution of wind energy as a Green Technology
2. Classify wind turbines into horizontal-axis and vertical-axis.
3. Describe various components of a wind energy system.
4. Compare onshore and offshore wind turbines in terms of feasibility and efficiency.
5. Discuss the importance of wind energy to mitigate the impacts of climate change.

4.1 Introduction

You must have played with small windmill toys in your childhood, which rotate with the help of the wind, bringing a cheerful joy to your face and keeping you bewildered as to why these toy-blades rotate with the help of the wind, correct? At an early age, you would have seen the big windmills rotating with the help of the wind to pump water from the Earth or to generate electrical power. Since ancient times, the power of the wind has been used for various traditional activities like water pumping and grinding. With the advancements in technologies and supportive policies from governments across the globe, wind energy is set to become a prominent player after solar energy in the global transition towards clean energy.

Wind energy is an indirect form of solar energy. Wind is caused by the movement of air masses due to uneven heating of the Earth's surface by the sun. The energy of the movement of air masses is captured by the rotational motion of windmills. This rotational energy is transferred through a shaft to an electrical generator. This mechanical energy is then converted to electrical energy with the help of generators. The basic principle involved is the aerodynamic force exerted by the wind on the blades to create lift and rotational movement like an airplane wing. Wind energy is an inexhaustible, abundant, and clean source of energy that can help us get rid of fossil fuel energy sources. Wind turbines may be installed offshore (sea-based) or onshore (land-based). Onshore wind farms are located in large water bodies, and they have the advantage of stronger and more consistent winds due to almost no obstructions.

Historical Development:

Wind energy has been in use for thousands of years. It has been on record that people in Persia (modern Iran) used wind-powered machines to grind grains and pump water around 500-900 AD. Windmills had become immensely popular across Europe in the 12th century. However, after the Industrial Revolution in the 18th and 19th centuries, there was a decline in the usage of wind energy as fossil fuels became the dominant source of energy due to easy availability and numerous applications that can be operated independently of weather conditions. In the 19th and 20th centuries, once again, wind energy found a place in the market with the introduction of small wind turbines to generate electrical power, particularly in rural areas.

Wind energy got big attention during the energy crisis of the 1970s, along with other renewable energy sources. Rising prices of oil, awareness towards environmental sustainability, energy security, and global warming have forced man to seek alternative clean fuels. Advanced technology and development in the fields of material

> ***Activity**: If our ancestors were able to prepare windmills in the dearth of tools, you must be able to prepare a wind plant.*
> *Design and prepare three distinct types of small wind power plants (Horizontal axis, Vertical axis, and Vibrating blades) to rotate the same small generator. Note the power generation in all cases.*
> *What did you observe? Can you think of a novel wind turbine design that can generate more power?*

science, power electronics, electrical generators, turbines, and integrated circuit technology have played significant roles in increasing the efficiency, cost reduction, and acceptability of wind power. Large wind farms are now deployed by the government as well as the private sector.

World Scenario:

Renewable energy sources like wind and solar play a significant role in curtailing the ill effects of global warming and climate change. To reduce the dependency on fossil fuels, almost all governments across the world are advocating for the deployment of large wind farms, which have contributed to a significant increase in the share of renewable energy in their energy mix. As of today, the total installed wind power plant capacity is approximately 1200 GW, which is 11% of global energy needs [17].

1. Leading Countries in Wind Energy Production:
   - China: China is the world's largest wind power producer with a fleet of onshore and offshore large wind farms.
   - USA: The United States is one of the major players in the field of wind energy. In the USA, Texas and Iowa are leading the sector.
   - Germany: Germany is a leader in offshore wind energy and has a clear plan for renewable energy.
   - India: India is aggressively adopting the wind sector to increase its stake in renewable energy. It has the advantage of vast coastlines with consistent wind speed. Gujarat and Tamil Nadu are the coastal states that use their vast coastlines, while Rajasthan has the advantage of the arid land of the desert. In India, wind energy contributes 44 GW, which is roughly 10% of its total installed power capacity as of 2024.
   - Denmark: Denmark is a pioneer in wind energy. According to publicly available data (of 2023), wind power constitutes 55% of the total power needs of Denmark. Denmark aims to achieve a benchmark of 100% renewable electricity by 2030, leveraging its geographical location. *(source: Danish Energy Agency)*

2. Growth Trends and Future Projections:

- As of 2024, the global cumulative wind energy capacity is estimated to be 1200 GW and is expected to cross 1800 GW by 2030 [17], with the Asia-Pacific region contributing more than half, Europe about a quarter, and America one-fifth.

- Offshore wind energy generation is also on rapid growth. Nearly 7 % of total wind energy generation is from offshore sites, particularly from Europe, China, and USA.
- Taller towers, larger motor rotor diameters, and advanced turbine technology are significantly contributing to higher efficiency and increased energy capture.

3. Impact on Economy and Employment:

- The wind energy sector supports jobs for millions of people worldwide through research, manufacturing, installation, and maintenance, ranging from high-skilled to semi-skilled and unskilled workers.
- The per-unit price of wind energy is constantly declining, making it compatible with fossil-fuel-based electricity generation. Thus, it is becoming a prime factor to positively contribute to the economies worldwide.

Importance of Wind Energy in Sustainable Development:

Wind energy plays a key role in achieving the Sustainable Development Goals (SDGs). It also addresses the burning critical questions of the present time, like climate change, energy security, and economic growth. Let us briefly discuss the environmental benefits, energy security, economic considerations, and challenges.

1. Environmental Benefits:
    - Wind energy is a clean energy source. It does not generate any greenhouse gases, such as CO_2 or other pollutants, which harm our environment. Thus, it helps to mitigate the climate challenge.
    - As non-renewable energy sources are not used, our natural resources are conserved. It simply uses the indirect form of solar energy.
    - Wind power plants do not require water for their operation. It is important to note here that thermal power plants and nuclear power plants require a huge amount of water for cooling, while hydropower plants can not function in the absence of water. Hence, wind energy is a water-efficient alternative.
2. Energy Security:

- Dependency on fossil fuels is reduced.
- It also leads to diversifying the energy mix.

3. Social and Economic advantages:
   - Wind projects are often located in rural and remote areas, leading to better employment opportunities in these areas.
   - This sector creates jobs in many areas at various skill levels in the areas of manufacturing, construction, operation, and maintenance.
   - Wind energy offers a decentralized grid, allowing better options to equip rural and remote areas with electricity.
   - Roof wind plants can help generate revenue as well as fulfill the need for electricity for an ordinary person, making them self-reliant and improving the quality of life.

4. Challenges:
   - Wind energy is intermittent. One can not predict its availability every time because it is completely dependent on the availability and speed of the wind. So, energy storage is required for backup generation.
   - Some people oppose the installation of wind power plants due to noise, visual impact, aesthetic concerns, and land concerns.
   - To reduce the impact on wildlife, careful site selection is required.

4.2 The Science of Wind Energy

Man has been harnessing wind energy for centuries for various tasks like grinding grains, pumping water from wells, and for fun. Modern technologies focus on generating electricity through wind energy. The wind is the movement of air masses. This movement of air is the result of differences in air pressure due to uneven heating of the Earth's surface.

4.2.1 Wind Basics

Since the output of the wind turbine is directly proportional to the cube of wind speed, even a slight increase in the wind speed can result in a significant increase in energy production because generated wind power is directly proportional to the cube of wind speed. The speed of wind depends on many parameters like its geographical location, altitude, and surroundings.

The following factors contribute to wind formation:

1. Solar Heating:
   Solar radiation heats the surface of the Earth unevenly due to its tilt, rotation, and the varying properties of land and water, including rivers, oceans, mountains, grasslands, deserts, residential areas, industrial areas, and forests. Also, it is affected by geographical location; for instance, polar regions will receive minimum solar radiation, while equatorial regions will receive maximum solar radiation. Due to this uneven heating, some regions become hotter and thus less dense, while some regions stay cooler and thus remain denser. Air masses in hot regions rise, creating low-pressure areas. Cool air from the surroundings moves towards low-pressure zones, creating an air movement called wind.
2. The Coriolis Effect: We have seen that in the Northern Hemisphere, winds are deflected toward the right, while in the Southern Hemisphere, winds are deflected toward the left. Think about why it is so. Is there anything related to the rotation of the Earth? Yes, it's due to the rotation of the Earth on its axis. It's due to the moment of inertia of air.
3. Local Wind Patterns: Local geographical features such as mountains, valleys, and grasslands play a vital role in forming local winds. They are localized areas and exist for certain periods.
   a. Sea and Land Breezes: During the day, land gets heated faster than water, leading to the movement of air from the sea towards land, forming a sea breeze. During the night, land cools quickly, but water bodies cool slowly, leading to the movement of air from land towards sea, forming a land breeze.

b. Mountain and Valley Winds: During the day, warm air rises in mountain regions, leading to Valley winds, while during the night, dense air flows down the slope, leading to Mountain winds.

4.2.2 Wind Turbine Basics

Wind is the movement of air. This moving air has kinetic energy. Wind turbines capture the kinetic energy of wind and convert it into rotational mechanical energy. This mechanical energy is then converted into electrical energy through the electrical generator. Wind turbines can be classified into two categories: Horizontal Wind Axis Turbines (HWAT) and Vertical Wind Axis Turbines (VAWT). HAWTs look like traditional windmills. VAWTs are used more in urban areas or areas with lower wind speeds due to their ability to function in variable winds. Almost all types of wind turbines consist of a tower, nacelle, blades, rotor, yaw system, and brake system. Now, let us discuss these components in detail.

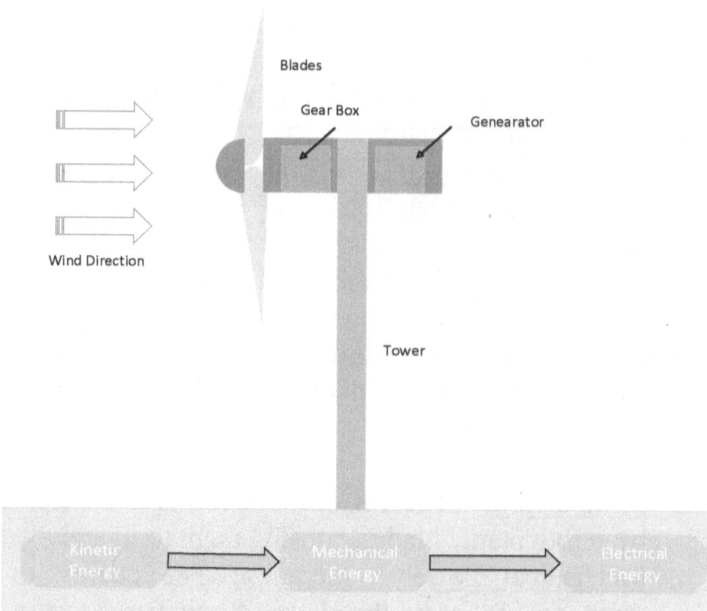

Fig.4.1 Components of Wind Turbine

Components of a Wind Turbine: A typical wind turbine consists of components like a rotor, blades, nacelle, tower, yaw system, and brake system.

a. Rotor and Blades: A rotor is a central axial part comprising a different number of blades. Blades are aerodynamically designed to get maximum efficiency using the aerofoil principle, similar to airplane wings. The flowing wind creates lift and rotation in the blades.
b. Nacelle: Nacelle houses the important components like Gearboxes, which convert the low rotational speed of the rotor into high speed, an Electrical generator, which converts rotational mechanical energy into electrical energy, and Controllers and sensors to regulate the turbine operation by wind conditions.
c. Tower: The Tower supports the rotor and nacelle. Its height may vary depending on the desired quality of wind required to have steady electricity generation. Towers are made of steel or concrete.
d. Yaw: Yaw helps to align the blades in the direction of the wind to achieve maximum electricity generation. It is provided in horizontal-axis wind turbines but is not required in vertical-axis turbines.
e. Brake Mechanism: The brake mechanism helps the turbine to stop operation in case of excessive wind speed or in case of cyclones to prevent damage. They may be operated in case of mechanical failure.

4.2.3 Working Principle of Wind Turbines

The following steps are involved in electricity generation through wind turbines:

a. Wind Capture: The very first action is to capture wind by aerodynamic blades, causing them to rotate.
b. Mechanical Power Generation: The energy of the rotational motion of the blades is transferred to the rotor and then to the gearbox.

c. Speed Enhancement: The required optimal speed is achieved with the help of a gearbox to suit electrical generators.
d. Electricity Generation: Turbines rotate electrical generators. These generators convert mechanical energy into electrical energy.
e. Grid Integration: Generated electricity is required to be transmitted to load centres through the central grid. Connecting wind power to the central grid is known as grid integration. Power conditioning is required to match the grid voltage, frequency, and other parameters. Meters are also installed to measure the amount of electricity wheeled into the grid.

4.2.4 Wind Energy Conversion Efficiency

According to the Betz limit, wind turbines can generate a maximum of 59.3% of the kinetic energy of wind. Higher than that would mean halting of wind, so no turbine can exceed this limit. Turbine design, tower height, and wind speed also affect the efficiency of wind turbines. Let us discuss these in detail:

a. Blade Design and Aerodynamics: Turbine blades should be aerodynamically designed to minimize air drag. They are manufactured with carbon fibre and fiberglass to increase durability.
b. Cut-in Speed: It is the minimum speed of wind at which electrical generation starts. It is generally 3 to 4 m/sec.
c. Cut-out Speed: It is the maximum speed of wind at which the turbine will stop rotating, to protect it from being damaged. Generally, its value used to be 25 m/sec.
d. Generator and Gearbox Efficiency: The Gearbox reduces efficiency, so now most wind turbines use direct drives. Advanced permanent magnet and superconducting-type generators are used to improve the efficiency.

e. Tower Height and Location: As said earlier, energy output is directly proportional to the cube of wind speed, so it is always desirable to place a turbine at an optimal height and proper location to capture the maximum wind. Offshore wind turbines, mostly located in the open sea, do not face any obstructions and hence offer higher efficiency.
f. Maintenance: Regular maintenance, using AI-based advanced preventive maintenance, advanced sensors, and real-time monitoring, can contribute to improving the efficiency and reliability of the system.

4.3 Types of Wind Energy Systems

Wind energy systems can be broadly classified into three categories. These are Onshore wind energy, Offshore wind energy, and Hybrid wind energy systems (Figure 4.2). In this section, we shall discuss the merits, demerits, challenges, and technological advancements for each system in detail.

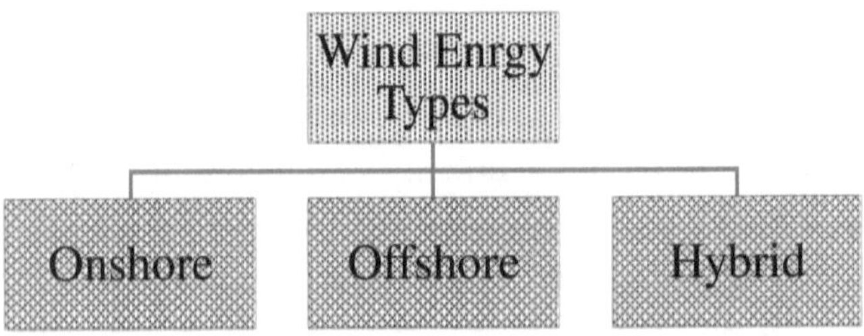

Figure 4.2 Types of Wind Energy

4.3.1 Onshore Wind Energy

Onshore wind energy refers to the wind farms that are installed on land. These are areas with high and consistent wind throughout the year. These machines may be installed in areas like the plains. Hills, seacoast, deserts, and valleys. These are the most common types of wind energy systems and are widely used across the globe for electricity generation.

Benefits of Onshore Wind Energy:

1. Since their sites are on land, their installations and transportation are always easy and less costly in comparison to offshore turbines.
2. Maintenance of land-based turbines is always easier due to easy access in all seasons.
3. Deployment of onshore wind turbines takes less time in comparison to offshore turbines.
4. Approval from various government departments is always easy in the case of onshore wind turbines.
5. Since it is proven technology, along with a well-established supply chain system, it has faster technological development and availability of spares.
6. Such projects create job opportunities for local economies.
7. New turbines can easily be added at any later stage.
8. Smaller wind turbines, even at the level of individual households, can lead to a distributed energy system.

Limitations of Onshore Wind Energy:

1. The onshore wind energy system requires a substantial chunk of land. This may compete with agricultural land, village expansion, or conservation areas.
2. Due to visual impact and noise pollution, many people oppose the installation of onshore wind energy systems.
3. Wind energy is highly unpredictable and intermittent. So, power generation is not reliable.
4. In case we desire reliability, then additional energy storage systems have to be added, which are very costly.
5. It may interact with the local ecosystem of birds and bats and may pose a risk to them.

4.3.2 Offshore Wind Energy

Offshore wind energy turbines are installed in large water bodies like oceans, rivers, or lakes. They have the advantage of stronger and more consistent wind because they do not have obstructions, thus making them more efficient than their onshore counterparts.

Advancements in Offshore Wind Technologies:

1. Traditional wind turbines are anchored to the seabed, so they need to be installed only in shallow water, while floating wind turbines are anchored to the seabed using mooring lines. So, it is possible to deploy them even into deeper seawater.
2. They can have a much higher capacity than offshore wind turbines, so a higher amount of electricity can be generated.
3. Now, durable and tougher materials are available that can generate power even in salty environments and harsh conditions for a longer time.
4. Bulk electrical power can be transmitted to load points using HVDC transmission lines very efficiently from remotely located wind farms.
5. Advanced AI and improved power electronics help to seamlessly transfer power.

Benefits of Offshore Wind Energy:

1. Since consistent and high-speed winds are available at offshore sites, they can generate more electricity.
2. Onshore windfarms have a capacity factor of 30-40%, while offshore windfarms have a capacity factor as high as 50% to 60%.
3. These are located on water bodies, so no land is required.
4. Since offshore wind farms are much farther from human settlements, there will be no concerns like noise pollution or visual impact.
5. Since there is no issue of land, offshore wind farms can be much larger.
6. Nearby coastal areas can benefit from offshore wind farms.

Environmental Considerations:

Offshore wind farms are remarkably successful and have many advantages, but it has many environmental impacts. The following are a few significant environmental considerations related to the deployment of offshore wind farms:

1. Offshore wind farms may disturb the patterns of fish migration and disrupt marine habitats. Hence, site selection

should be done very carefully to minimize the environmental impact.

2. Marine mammals, migratory birds, and bats can be affected by the noise produced by the turbine blades.
3. It should not be in fishing zones or shipping routes.

Despite these challenges, offshore wind farms are becoming a popular choice in the renewable energy mix at the global level, particularly for countries with large coastlines.

4.3.3 Hybrid Wind Energy

In this case, wind energy is integrated with other modes of energy generation, like solar, natural gas, or conventional fossil fuels. Since wind energy is intermittent, this hybrid mechanism ensures continuous energy output.

1. Wind-Solar Hybrid System: Wind and solar energy are renewable forms of energy. Solar energy is abundant during the daytime but not available during the night. Wind speeds are low during the daytime but higher during the night. So, wind and solar energy complement each other. Hybrid systems share many infrastructures like land, batteries, and inverters, which reduces the overall cost.
2. Wind-Battery Storage System: As wind energy is intermittent, it may happen that sometimes wind energy is not available. To get rid of this problem, batteries are used to store the excess power and are used when wind power is not available or is only partially available.
3. Wind-Diesel Hybrid System: This system is used for remote areas where grid access is not available. In case of low wind power generation, diesel generators are used.
4. Wind-Hydro Hybrid System: Pumped hydro storage power plants can be integrated with wind power plants to ensure the continuity of electrical power. Excess wind power is used to pump water to a higher reservoir to generate electricity during low wind speeds.

Benefits of Hybrid Wind Systems:
It is well known that winds are highly unpredictable, and so is their speed; hence, electrical power output can not be relied upon from wind power plants. Hybrid systems increase the stability of the system, reducing the intermittency and fluctuations of wind power plants. Maximum utilization of wind energy is achieved, and fossil fuel is only used in case of unavailability of wind, hence it reduces the greenhouse gas emissions.

Challenges of Hybrid Wind Systems:
Hybrid systems often require high initial investments due to additional components for energy storage and others. Management of multiple energy sources requires complex mechanisms and control strategies. Integration with the grid also requires advanced machinery. Its use on a large scale has not yet been evaluated and needs much research. Hybrid energy systems are always difficult to manage and maintain, and the technology related to them is still evolving. Hybrid wind energy cleverly solves the intermittency issues of wind power. Seeing the advantages of reliability and efficiency, hybrid wind energy systems will play a significant role in the renewable energy sector.

4.4 Wind Energy and Environmental Impact

Wind energy is indeed a clean source of energy which do not generate any greenhouse gases and thus does not pollute our environment, but wind energy has certain environmental impacts that must be carefully studied before its deployment. Its effect on soil life, noise pollution, the life cycle of its components, and its impact on aesthetic look are significant factors worth discussing.

1. Reduction in Carbon Footprint: Wind energy is the cleanest source of energy, having a minimum carbon footprint even during its manufacturing. Approximately 2000-3000 MWh of electricity can be generated using a 1 MW capacity turbine every year, which can help to satisfy the electricity demand of approximately 1000 homes! This is equivalent to a reduction of ~2500 metric tons of CO_2 if the same power is generated through fossil fuels. According to

celanpower.org, global wind power helps to avoid approximately 351 million metric tons of CO_2 annually (as of 2023).

Energy Payback Period: The energy Payback period is the time required to generate as much energy as was used in the production of that machine. Like other equipment, wind turbines also require energy for their manufacturing, transportation, and installation, but wind turbines very quickly offset this carbon footprint. Depending on the size and location of the wind turbine, the energy payback period is usually 3 to 9 months. After this initial payback period, Wind turbines provide carbon-free electricity for the rest of 25-30 years.

2. Impacts on Wildlife: The major environmental concern often raised against wind energy is its impact on birds and bats. If wind turbines are installed on the route of migratory birds, then they may obstruct their way, so careful selection of the site is necessary. Birds and bats may accidentally collide with the blades of wind turbines. Wind farm developers should plan habitat protection programs to offset the losses. Bats are more prone to colliding with turbine blades than birds. Wind farms near forests pose higher risks for the bats. This can be mitigated through various strategies to deter the bats from approaching turbine blades by emitting high-frequency sounds.

   Offshore wind turbines are beneficial as they do not require land and have steady and consistent output power, but they have a certain environmental impact on marine life. Turbine installation generates underwater noise that may disturb whales and dolphins. It is also seen that certain fish species have changed their habitat due to disturbances of electromagnetic fields generated by marine cables. These challenges can be mitigated by air bubble barriers around turbine foundations to reduce noise and by seasonal construction timing to avoid the breeding period of marine animals.

3. Noise and Aesthetic Concerns: Wind turbines produce two types of noise. Aerodynamic Noise is generated by the movement of blades, usually a "whooshing " sound, and the other one is Mechanical noise generated from the gearbox and generator. However, most modern turbines use direct drive technology, which eliminates the need for gearboxes. To reduce the noise concern, most governments across the globe have regulations that do not allow the deployment of wind turbines within a radius of 1000 meters from residential areas.

   Some people oppose wind farms due to their feelings that wind farms disrupt natural beauty and landscapes, especially in coastal areas. However, proper community engagement may address this challenge very peacefully. Certain other programs, like involving local people in the revenue sharing, priority of local people in employment opportunities, and underground cables to maintain the natural beauty, can also play a significant role in convincing people around the wind farms. Despite these concerns, it is observed that most people support wind energy, realizing its environmental benefits and economic contributions.
4. Lifetime Environmental Impact of Wind Turbines: For any machine, it is important to have the Life Cycle Assessment (LCA) to evaluate its impact from production to decommissioning. The following are the stages of wind turbines in their lifespan:
   a. Manufacturing and Material Use: Wind turbines are made of steel (~70-80%) for the tower, foundation, and other fabrications; fiberglass for the blades; and certain other metals like copper for generators. Cables and transmission lines are made of copper, steel, plastic, and other materials. The mining and material extraction process is required to get these minerals, which contribute to carbon emissions and resource depletion.

b. Transportation and Installation: Wind turbines, generators, and their blades are often big and heavy, and heavy vehicles are required to move them to the installation site. Installation also requires heavy cranes, sometimes helicopters, which generate carbon emissions.
c. Operation and Maintenance: Wind turbines have a lifespan of 20 to 30 years. Wind power plants are also less complex and hence require less maintenance in comparison to fossil fuel-based power plants. They produce far more energy than they consume during their life cycle.
d. Decommissioning and Recycling: The majority of the components of a wind turbine can be recycled. However, it is difficult to recycle wind turbine blades because they are made of composite fibres, but they may be repurposed in the construction of bridges, thereby improving sustainability.

The following is a comparison of the carbon footprints of various power plants.

- Wind Power: 10-12 g CO_2/kWh
- Solar Power:40-50 g CO_2/kWh
- Natural Gas:400-500 g CO_2/kWh
- Coal Power:800-1000 g CO_2/kWh

From the above comparison, it is noticeably clear that wind energy is the cleanest source of energy with the lowest environmental impact. With improved site surveys, better turbine designs, and responsible life cycle management, wind energy can become the prime choice as a sustainable and eco-friendly energy source.

4.5 Technological Advancements in Wind Energy

With the help of technological developments, improved efficiency, reliability, longer life, minimum maintenance, and being the cleanest source of energy, the wind energy sector is rapidly evolving. In this section, we shall discuss innovative turbine designs, floating wind turbines, smart grid integration, and emerging materials for turbines.

1. Innovative Turbine Designs: Modern turbines are large, capable of generating 15 to 20 MW per unit. These large turbines can harness more wind energy per unit, which reduces the number of turbines per wind farm. As of 2024, the world's largest wind turbine is a 20 MW offshore turbine developed by Mingyang Smart Energy. It has a rotor diameter of 260-292 meters [31]. The Goldwing GWH252-16MW turbine installed in the Fujian Province of China is the largest physical wind turbine. Each blade of this turbine measures more than 123 meters! [32]

   Unlike traditional horizontal-axis wind turbines, VAWTs have a vertical axis, which protects them from air turbulence, allowing them to work more efficiently. They can operate in urban environments with complex wind patterns. They require less space in comparison to traditional horizontal-axis turbines. Most components are located near the ground, which offers easy maintenance.

   Bladeless wind turbines, such as Vortex Bladeless, do not have rotating blades but vibrate to generate electricity. Thus, they have almost no noise, minimal maintenance, and no threat of bird and bat collisions. They are more compact, making them suitable for residential use in urban and suburban areas.

2. Floating Wind Turbines: Most wind farms are located near seashores in shallow water to grout the base of the turbine on the seabed, but in reality, 80% of global offshore wind resources are available in the deep sea, which requires floating wind turbines. The depth of the sea may be more than 60 meters, where it would be impractical to anchor a foundation on the seabed. Floating wind turbines are in deeper waters where we find strong and consistent winds. As of June 2025, Hywind Wind Farm, Norway, is the

world's largest floating wind farm, with a capacity of 88 MW [33]. Japan, France, Spain, and USA are also exploring this field. Floating wind turbines have the advantage of increased energy output due to strong, consistent, obstacle-free, vast areas; they do not compete with limited land area, and they do not pose any challenges of noise and other environmental degradation.

Despite floating wind power plants offering many advantages, they do have certain limitations. Since they are located far from the seashore, it is difficult to transmit power to the consumers. They require a special class of advanced cable for efficient power. Seeing the potential of floating wind turbines, countries like Norway, Japan, and the USA are heavily investing in research and development of floating wind turbine technology.

> ***Thinking Corner***: *Don't you think floating wind power plants will sail away in the sea due to the winds? What's the solution to keep it stable and stationary in the deep sea?*

3. Smart Grid Integration and Storage Solution: Since wind energy is highly intermittent and can not be predicted perfectly, its continuity of supply can not be assured. This requires an advanced grid solution to balance power supply and demand. Better grid management is possible through real-time digital modelling of wind farms. The demand response system should adjust the power consumption based on the availability of wind energy. AI-powered weather prediction systems should be implemented to optimize wind farm operations. Battery storage (Lithium-Ion, Flow batteries, Solid state batteries) should be utilized to fulfill the peak load. Tesla Megapack and Flow batteries offer longer-duration energy storage and are suitable for wind integration. Surplus wind energy can also be used to pump water back to higher reservoirs and can be used to meet the peak hours demand. In the same way, excess wind energy can also be used to produce

Green Hydrogen using electrolysis. This Green Hydrogen can be used for transportation, in industries, and in fuel cells. Many countries like Japan, Germany, and Australia are investing in large-scale wind-powered Green Hydrogen Production Plants.

4. Emerging Materials for Turbines: As mentioned at the beginning of this chapter, our ancestors were very well acquainted with wind energy, and they harnessed the wind energy for various tasks like grinding grains and pumping water from wells, but wind energy could not be tapped beyond a certain level due to a lack of proper materials for its manufacturing. To make wind energy economically viable and on a large scale, bigger and bigger wind turbines need to be manufactured. With the advancements in technology and research in metallurgy, stronger, lighter, more durable, and anticorrosion materials are available, which have increased the efficiency of wind turbines. Carbon fibre and Composite Blades are much lighter and stronger than traditional fiberglass blades. It increases its energy capture efficiency and reduces wear. Traditional fiberglass materials were difficult to recycle, but companies like Siemens Gamesa have developed fully recyclable blades that reduce waste. Certain companies like GE and Vestas are using the latest technology of 3D printers to make certain turbine parts to reduce the cost of manufacturing and increase aerodynamics. To reduce the environmental impact, plant-based resins are being used to prepare biodegradable wind turbine components.
5. Artificial Intelligence and Big Data in Wind Energy: Artificial intelligence has transformed the lives of human beings. It has also revolutionized the wind energy sector by enhancing performance, predictive maintenance, and grid integration. AI is helping to predict the weather conditions, thus reducing the grid fluctuations. Machine learning can be employed to increase the profitability of wind farms. AI-assisted predictive maintenance can detect faults before they occur through data acquired from sensors placed in turbines. This reduces downtime and repair costs. AI-powered drones can inspect the turbine blades for any wear

and tear, thus reducing maintenance time. The AI-based algorithm helps to stabilize energy distribution based on electricity demand and grid conditions. In the case of hybrid wind power plants, AI-based systems can maximize the use of wind energy with minimum use of parallel plants (like fossil fuel-based) to meet the varying demand during varying wind output.

4.6 Wind Energy in a Global Context

Seeing the dark side of global warming, climatic calamities, and ever-increasing demand for sustainable sources of energy, the world has no option but to research, invest, and implement renewable energy sources. Wind energy is the cleanest form of energy, with minimal environmental impact. As of 2023, the cumulative installed wind power across the world is estimated to be 1200 GW. This is 11% of our global electricity needs. Globally, 116 GW of new capacity was added in just one year. This is the highest annual addition ever! [17]. The following figures 4.3 and 4.4 show the growth in cumulative wind power capacity worldwide and in India, respectively.

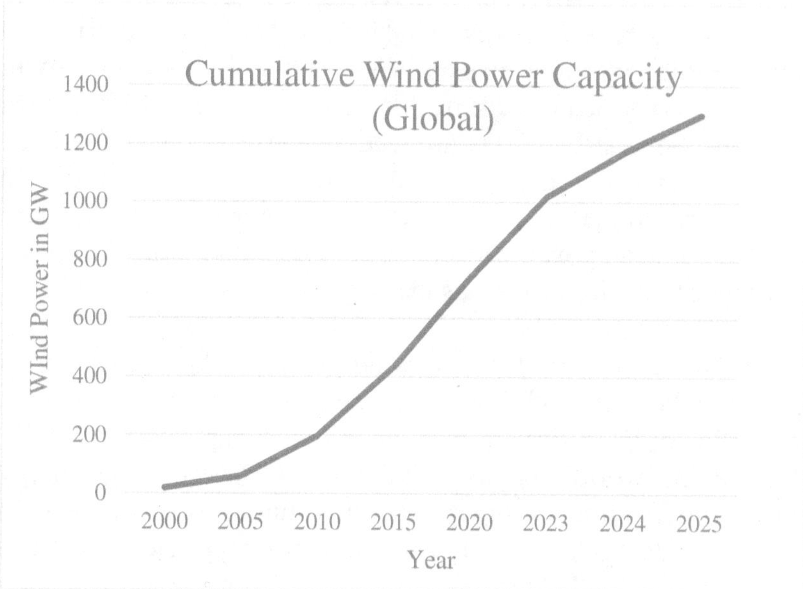

Figure. 4.3 Global Cumulative Wind Power Capacity in GW [17]

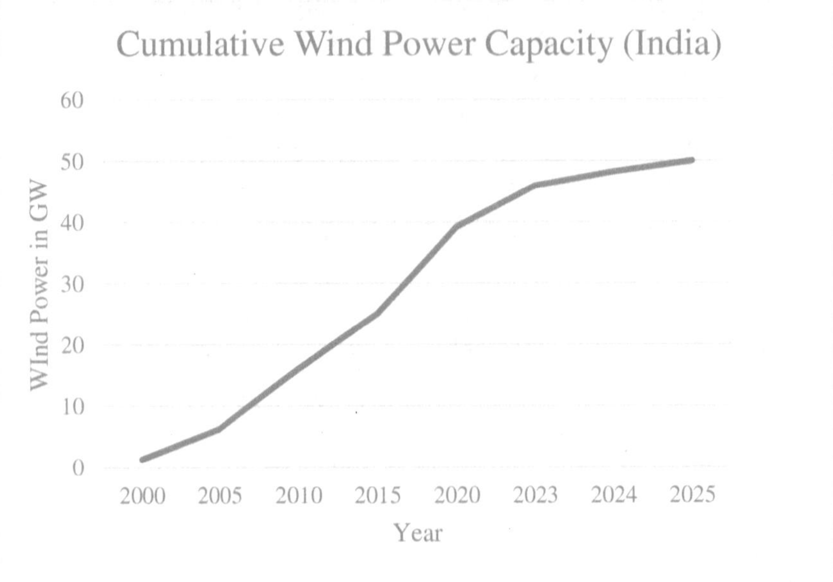

Figure 4.4 Cumulative Wind Power Capacity of India. [4]

4.6.1 Leading countries in wind energy development

Unlike other sources of energy, wind is available in all countries and locations with varying speeds and consistency. Every country will try to tap into and avail itself of this easy option to fulfill the need for energy. Several countries are leading the way in wind energy generation and research. These nations have invested in infrastructure, framed positive policies, and supported the industry to expand the wind power capacity. The following Figure 4.5 shows the installed capacity of various leading countries.

China is the largest producer of wind power. It has expanded its capacity in both the offshore and onshore wind energy sectors. Jiuquan Power Base aims to have an installed capacity of 20 GW, making it the world's largest wind farm. Offshore wind energy plants with floating turbines are also under development in Guangdong, Jiangsu, and Fujian. Motivating policies and easy regulations have fuelled the rapid expansion. China plans to increase its wind power capacity to 450 GW by the end of this decade [17].

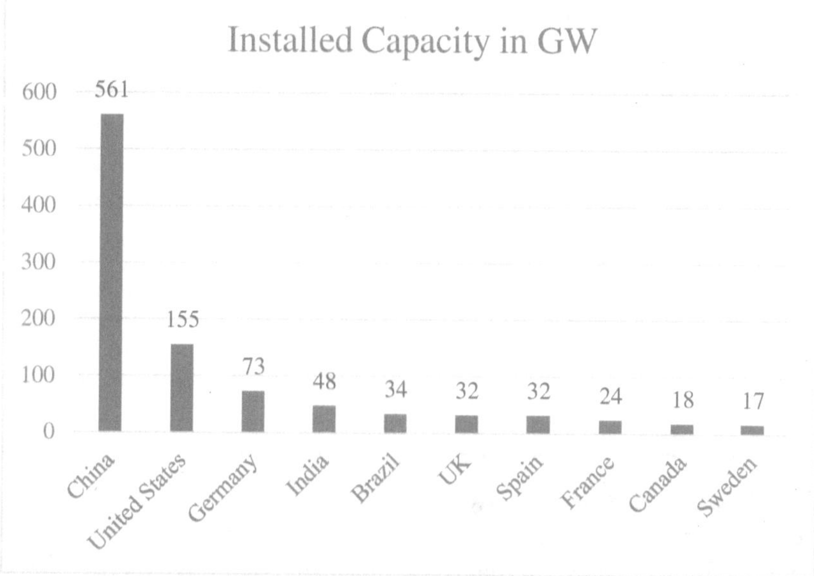

Fig. 4.5 Country-wise installed wind power capacity
Source: Data compiled from World Population Review, Wikipedia, and other publicly available reports, February 2025

The United States of America is the second-largest wind energy producer. Major sites are located in Texas, Iowa, and Oklahoma. It may surprise readers to know that Texas alone generates more than 40 GW, which is more than most countries. The Atlantic and Pacific coasts are seeing new developments in offshore wind power projects. The USA has outlined a goal to generate 20% of its nation's demand from wind energy by 2030.

Germany has been a pioneer in wind energy and is a global leader in offshore wind energy. It has developed strong policies and frameworks to support offshore wind farms. The Energiewende policy has made wind energy the largest electricity source. Germany is planning to have 100% renewable energy by 2045. The North Sea Offshore wind farms are going to play a significant role in this mission.

India is one of the fastest-growing energy markets in the world, with an emphasis on renewable energy sources. India has been bestowed a large coastline by nature, with steady and consistent winds. Tamil Nadu, Rajasthan, Gujarat, and Maharashtra are playing a significant role in achieving the country's target of 140 GW by 2030. Hybrid energy plants are also popular in India.

Brazil, Spain, France, Canada, Denmark, Sweden, and many other countries have targets to expand their renewable energy sources with wind energy.

> ***Activity**: List a few potential wind sites near your college and home, where you observe continuous strong winds. Can you name a few important wind power plant sites in India and your state?*

4.6.2 Policies and Regulations

Government policies are there to regulate the installation of wind turbines. It has been seen that those countries that have well-defined, strong policy frameworks have seen rapid growth in the wind sector. In this section, we shall elaborate on the role of government policies and their impact.

a. Feed-in-Tariffs (FiTs): In this system, pre-negotiated fixed prices are guaranteed for the generated electricity. Due to the assurance of a fixed price, such a policy encourages long-term investment. Germany, China, and Spain are using this policy.
b. Renewable Portfolio Standards (RPS) and Mandates: In such a system, utilities are required to generate a percentage of their energy mandatorily from renewable sources. Through the Texas RPS policy, Texas has seen a boom in wind energy. India and the USA follow this policy.
c. Auctions and Competitive Bidding: Many countries, including India and Brazil, prefer auctions and bidding to get wind energy at the lowest possible price. This has brought down the price of wind energy production.
d. Tax Incentives and Subsidies: Low-interest loans and subsidies attract investors to establish wind farms on a large scale. Subsidies on renewable energy sources are never a

burden but a patch to repair the broken sheet of our climate, so it is always cheaper in the long term. The USA and China are showing the way to the world through such policies.

e. Offshore wind regulations: Land is a limited, precious resource; hence, many countries are motivating investment in offshore wind farm projects. Japan and South Korea have created floating wind policies to expand their offshore capacity, whereas the European Union and the United Kingdom have streamlined their license system to accelerate offshore wind farm installations.

> ***Thinking Corner****: Pause reading of this book here. Download the wind power policy of India. Do you feel any incentives/motivations to become an entrepreneur in this sector? It's an ocean of opportunities. Extract the areas of your interest to start a business or community service.*

4.6.3 International Agreements and Cooperation

To reduce global carbon emissions, global warming, and accelerate the growth of renewable energy worldwide, international agreements and cooperation are important. It helps underdeveloped countries acquire technology and financial support from developed countries. The following are the key international policies to support wind energy development:

a. Paris Agreement: This agreement forces nations to reduce greenhouse gas emissions and promote the use of renewable energy sources like wind energy. Wind energy plays a prominent role in achieving the net-zero carbon target by 2050 as pledged by the Paris Agreement.
b. COP Climate Conferences: COP26 (Glasgow) held in 2021 insisted all member countries relinquish the use of fossil fuels and expand renewable energy sources like wind to fulfill the need for energy demand. COP28 (Dubai), held in 2023, insisted on tripling the global renewable capacity by 2030.
c. Global Wind Energy Council (GWEC) Initiatives: It promotes wind energy through technology collaboration,

market research, and policy advocacy. Its programs, like the "Offshore Wind Accelerator," helped to boost wind installations in emerging markets [17].

d. EU Green Deal and REPowerEU plan: The EU Green Deal is a comprehensive strategy by the European Union. The EU Green Deal aims to achieve zero carbon neutrality by 2050 by reducing greenhouse gas emissions. The REPowerEU plan is a part of the EU Green Deal to reduce fossil fuel dependence through diversifying energy sources [10].
e. Asia-Pacific Wind Energy Collaboration: Floating wind power is expected to be a major growth area in the Asia-Pacific region. China, South Korea, and Japan are working together in this field to share technology and innovations.

4.6.4 Successful Wind Energy Projects

In the last few decades, there has been a surge in the deployment of wind farms worldwide. The following are a few important projects:

1. Hornsea Windfarm: It is in the United Kingdom. It is the world's largest offshore wind farm. It is situated in the North Sea of the UK with a capacity of 2.8 GW, which is equivalent to fulfilling the needs of over 2.5 million homes. It uses 13 MW giant turbines from Siemens Gamesa. It is also connected to the UK's smart grid for efficient power distribution.
2. Texas Wind Energy Boom: It is not a single wind farm. Major projects across Texas state, USA are the Roscoe wind farm and Horse Hollow wind energy centre. Alone, Texas supplies more than 40 GW of installed capacity. The main reasons are the availability of land and good wind speed, along with policy support from Texas.
3. Jiuquan Wind Powerbase: It is located in China and is the world's largest onshore wind farm, having a planned capacity of 20 GW, out of which 10 GW is already installed. Since it is situated in a remote area, the Government of China is investing in Ultra High Voltage (UHV) transmission lines to transport the generated power to deliver to cities.

4. Gujarat's Wind-Solar Hybrid Park: It is situated in Gujarat, India. It is the world's largest renewable energy park involving a hybrid mode using wind and solar energy. Its capacity is more than 30 GW. Its key features include hybrid energy storage and smart grid integration. It incorporated the bidding system to reduce costs.

4.7 Economic Aspects of Wind Energy

Wind energy has become one of the most competitive and widely used alternative options worldwide. Its deployment affects many economic aspects like investment trends, economic growth, job creation, skill enhancement centres, and financial mechanisms. In this section, we shall try to explore the cost of wind energy production, employment opportunities, and its comparative analysis with other renewable energy sources.

4.7.1 Cost of Wind Energy Production

It has been observed that the per-unit cost of wind energy production has significantly dropped during the last three decades due to several factors. This reduced cost has made wind energy an affordable and economically viable alternative to many energy sources, including traditional fossil fuels. The most common metric used to compare various energy sources is the Levelized Cost of Energy (LCoE). The following table represents the trend of declining cost per MWh of wind energy using LCoE.

Year	Price per MWh for	
	Onshore windfarm	Offshore windfarm
2010	Rs 6400 – Rs 8000	Rs 12000 – Rs 16000
2020	Rs 2400 -Rs 4000	Rs 6400 – Rs 9600
2023	Rs 2000 -Rs 3600	Rs 4000 – Rs 6000

Table 4.1 Price per MWh for Onshore and Offshore wind farms.

Due to the ease of maintenance and the availability of low-cost desert and arid regions, the price per MWh for onshore wind energy has declined to as low as Rs 2000. The main factors affecting the price of wind energy are:

a. Capital Expenditure (CAPEX):

Onshore Wind: Rs 96 - Rs 120 million per MW

Offshore Wind: Rs 200 - Rs 360 million per MW (higher due to installation and transmission costs)

b. Operational and Maintenance (O&M) costs: Maintenance will be required for every machine. Since wind turbines have rotating parts, they will require more maintenance in comparison to PV solar energy. However, in comparison to traditional fossil fuel energy sources, wind farms require low O&M costs, typically ranging between Rs 800 and Rs 1200/MWh. Predictive maintenance has reduced costs.
c. Capacity Factor and Efficiency: It refers to the average power generation compared to the maximum power that can be generated. Onshore wind turbine machines have capacity factors of about 35 -50%, whereas offshore wind turbine machines have capacity factors of about 50-60%. This is because offshore wind turbines have stronger and consistent wind availability.
d. Parity with fossil fuels: Traditionally, fossil fuels have been the primary source of energy. Compatible infrastructure and technology to harness the energy from fossil fuels are already available, so energy generation from fossil fuels is much cheaper. In a couple of years, the breakeven point of wind energy will have been much reduced, making it economically viable compared to traditional fossil fuels. Wind energy is now cheaper than coal and natural gas energy in many regions of the world, as it requires only an initial investment and no running costs due to free fuel (wind). On the other hand, the price of fossil fuels keeps fluctuating due to world political events, wars, transportation issues, and sometimes scarcity.

4.7.2 Investment in Wind Energy: Wind energy requires a big initial capital investment. Such a big fund allocation requires the involvement of public and private sources. According to the International Energy Agency (IEA) report 2023, investment in the wind energy sector increased by 20%, reaching a record of Rs14,400 billion [3]. According to a World Wind Energy Association (WWEA) report, a record 116,065 MW capacity wind turbines were added globally in 2023 [34]. The following are the key financing mechanisms:

- Government Incentives and Subsidies: Production Tax Credit (PTC) provides tax benefits for wind power producers. Feed-in Tariffs (FiTs) guarantee a fixed price per unit of electricity generated, etc.
- Private Investments and Green Bonds: Bonds are issued to raise funds. In 2023, Rs 4000 billion was raised through Green Bonds. Many institutional investors like BlackRock and Brookfield also fund large-scale wind farms.
- Public-private partnership: Large-scale windfarms and offshore windfarms often rely on PPPs due to the need for heavy investments. The Dogger Bank wind farm of the UK and the Hornsea Project may be cited as examples.
- Corporate Power Purchase Agreements (PPAs): Giant companies used to buy wind power through PPAs to fulfill their mandatory sustainability goals as decided by the governments. Globally, it is estimated that big corporations have signed PPAs for over 100 GW of wind energy.

4.7.3 Job Creation and Economic Growth:

Wind energy can provide employment and contribute to economic development worldwide. The following is a rough estimate of jobs created through this sector.

Sr No.	Region	Jobs in the wind energy sector (2023)
1	Global	14,00,000
2	China	5,50,000
3	United States	1,20,000
4	European Union	3,00,000
5	India	65,000

Table 4.2 Global Jobs Opportunities in the Wind Sector.

Types of Jobs Created in the Wind Energy Sector:

- Manufacturing: To produce turbines, blades, and other mechanical components
- Electronics Industries: To produce PCBs, Electronic Circuits, Power electronics components, Computers, relays, sensors, Software, etc.
- Construction and Installation: Building the foundation of onshore and offshore wind turbines, Erection of wind turbines and towers.
- Electrical Sector: Manufacturing of power transmission cables and related accessories, Motors, Generators, Smart Grid related works, Electrical Substations, Maintenance of electrical items
- Operation and Maintenance: To ensure maximum efficiency.
- Transportation and Logistics
- Research and Development for innovative turbine designs, grid integration, etc

Economic growth is a by-product of job opportunities. Texas's wind boom has generated more than 30,000 jobs, while India's wind expansion has also generated thousands of jobs directly. Denmark's wind energy sector contributes more than 6% of its GDP.

4.7.4 Comparative Analysis between Wind Energy and other Renewable Energy sources

Sr No	Parameter	Wind Energy	Solar Energy	Hydro Power	Geothermal Energy
1	Capacity factor	35% -50 % (onshore)	20% - 25%	40% -60%	70% -90%
2	LCOE in Rs /MWh	Rs 2000- Rs 3600 (onshore)	Rs 1400 – Rs 2000	Rs 2000- Rs 6400	Rs 4000- Rs5000
3	Impact on the environment	Low (wildlife concerns)	Low (recycling issues)	High (ecosystem disruption)	Low

4	Availability	Day and Night	During the Days only	Day and Night	Day and Night
5	Land use	Moderate	High	High (dams)	Low
6	Scalability	High	Very High	Medium	Low

Table 4.3 Comparative Analysis of Various Renewable Energy Sources.

Extract of Comparison:

- Wind vs Solar: Wind farms have higher efficiency and can generate energy even during the night. Solar power is cheaper on a small scale. Wind power plants require less land in comparison to solar power plants.
- Wind vs. hydropower: Wind farms are easier to install and expand. Hydropower plants have a higher capacity factor but can disturb river ecosystems.
- Wind Vs Geothermal: Geothermal energy has higher efficiency and even the advantage of providing heat in colder regions, but it is available in limited and specific locations, while wind has wider global availability.

Wind and Solar energy are twin sisters that complement each other. Solar energy is available during the day, but is fully unavailable during the night, while wind energy is partially available during the day, but is more available during the night. Thus, they can complement each other to balance each other's intermittency through Hybrid Solar-Wind farms. India, China, and the USA are leading players in hybrid renewable energy projects.

4.8 Challenges and Limitations of Wind Energy

It is well-accepted and understood that wind power is one of the fastest-growing and cleanest sources of energy, but there are certain challenges and limitations in its deployment. These challenges and limitations need to be addressed for its widespread adoption. They may be linked to geographic or climatic limitations, intermittency, storage issues, public perception, and policy hurdles. Let us briefly discuss these challenges in this section.

1. Geographical and Climatic Limitations: Wind is available across the whole globe, but it becomes economically viable to harness if winds are blowing consistently at sufficient speeds for a certain period in a year. So, its effectiveness and implementation depend upon the location and climatic conditions of the area. Wind turbines require consistent wind speeds between 5 m/s and 25m/s. Seasons also affect its availability. Hurricanes and storms can damage the wind blades and turbines. Cold climates like Canada and Scandinavia face icing issues.
2. Land Constraints: Onshore wind farms require a large area, which is not possible in urban areas or places where land costs are too high. Mounted and forested areas are also not suitable due to wind turbulence.
3. Intermittency: Wind power is unreliable and unpredictable, as well as variable in nature. This creates challenges for grid stability and energy reliability.
4. Energy Storage: Since wind power is inherently intermittent, large-scale battery storage is required, which is expensive. Pumped hydro storage may be effective, but it is available only in mountainous areas. Hydrogen storage is a promising alternative, but it is in the research stage and is also costly.
5. Grid Integration Issues: The power grid faces the issue of balancing the intermittent variable wind power. Offshore wind farms and many onshore wind farms are also located in remote areas, which require long transmission line networks to extract the power. HVDC lines are required to transmit this power efficiently.
6. Public Perception (NIMBY Syndrome): Despite wind energy offering many advantages, wind projects are often delayed due to many reasons. One of the major reasons is NIMY (Not in My Backyard (NIMBY)

syndrome, which refers to the opposition of locals due to aesthetic, environmental, and noise issues.

7. Policy and Regulatory Hurdles also often cause delays or sometimes even lead to the cancellation of wind farm projects. Frequent changes in the policies related to subsidies, tax credits, tariffs, licenses, and regulations demotivate investors. Some government also favors fossil fuels due to their low price, abundance, and easy availability in their country, for example, Venezuela and Gulf countries, where plenty of oil is available at a cheap rate.
8. International supply chain systems and international geopolitical issues also play a vital role in wind farm project delays. Wind turbines and other components depend on global supply chains, which are affected by trade policies. For example, as of March 2025, tariff wars between the US and Canada and the US and China are affecting the price of many components and materials.

4.9 Conclusion

Our world is facing the impacts of global warming and climate change, and urgently needs to shift to renewable energy sources to achieve sustainable development. Wind energy is the cleanest form of energy, having minimal environmental impact with an extremely low carbon footprint in its manufacturing. Unlike fossil fuels, wind power plants do not produce greenhouse gases. Thus, they are environmentally friendly.

Wind energy has been known to man for centuries, and it was harnessed for grinding grains and pumping water from wells. Advanced technology and research have increased the efficiency and scale of wind power plants from kW to MW. Now, even GW wind farms are also installed worldwide to tap the maximum possible wind energy. China, India, USA, and Germany are leading countries in harnessing wind energy.

Worldwide, wind energy contributes nearly 10 % of global energy demand.

The wind energy sector also significantly contributes to economic growth and generates job opportunities in the manufacturing, installation, maintenance, and infrastructure sectors. Continuous research has helped to design more efficient turbine designs; however, challenges of intermittency and unpredictable wind patterns remain the major challenges for deploying wind energy. Other challenges include energy storage, impacts on wildlife, noise pollution, aesthetic concerns, and land requirements. High initial investment and government policies are a few other deterring factors. However, decreasing the cost per unit and the long life of wind turbines will help the wind sector enhance its share in the global energy mix.

To summarize, we may say that wind energy plays a very important role in sustainable development. It not only provides clean energy but also helps to address the global climate challenge. Through wind power, we move closer to our target of a green and prosperous future for the next generations.

> ***Thinking Corner**: Suppose a company is planning to install a windmill near your house. To feel the situation of continuous noise (from the perspective of people living in the vicinity of wind power plants), you downloaded the whooshing sound of a windmill from the internet and listened to it for an hour while doing your routine work.*
>
> *What should you do?*
> *(a) Accept this noise for the sake of sustainability.*
> *(b) Oppose the wind plants*
> *(c) Sell the house and go away*
> *(d) Request the government to locate the plant 1 km away.*

-:EXERCISE:-

Section A: Short Answer Questions

1. Wind power plants meet............. % of global energy needs.
2. Name two leading states in India for wind energy production.

3. Name five leading countries for wind energy production.
4. Explain the Betz limit in the wind energy conversion efficiency.
5. State True or False. “Offshore wind energy is considered more efficient than onshore wind energy.”
6. State the role of the yaw system in a wind turbine.
7. List three broad categories in wind energy systems.
8. Recall one major environmental concern related to offshore wind farms.
9. .... is the typical payback period for a wind turbine.
10. Discuss the role of a wind-solar hybrid system in energy generation.

Section B: Long Answer Questions

1. Explain the role of uneven heating in the formation of wind.
2. Compare wind energy with solar energy.
3. Compare wind energy with traditional fossil fuel power plants.
4. Draw and explain HAWT along with the function of major components.
5. Discuss the challenges in installing large wind farms.
6. Explain the role of wind energy in fulfilling the UN SDGs.
7. List limitations of wind energy.
8. Describe the innovations in wind turbine designs.
9. Analyse the role of hybrid wind plants in ensuring a reliable energy supply.
10. Prepare an expenditure statement to install a 3kW wind power plant on your terrace.
11. Explain the effects of offshore wind energy on marine life.

Section C: Multiple Choice Q and A

1. The wind power plants supply .... of global energy demand:
   a) 16 % b) 11% c) 23% d) 42%
2. The ... country has the largest installed wind power capacity.
   a) India b) Japan c) China d) France
3. Two states of USA leading in wind energy production are:
   a) Texas and Iowa b) Alaska and Montana
   c) New York and Ohio d) California and Florida

4. Two Indian states using vast coastlines for wind energy are:
   a) Gujarat and Tamil Nadu
   b) Odisha and West Bengal
   c) Kerala and Maharashtra
   d) Andhra Pradesh and Kerala
5. The installed wind power capacity of India as of 2025 is:
   a) 25 GW b) 44 GW c) 14 GW d) 104 GW
6. Denmark's......... % of total power needs are met by wind power:
   a) 10% b) 45% c) 55% d) 75%
7. By the end of this decade, global wind power is expected :
   a) 1000 GW b) 1800 GW c) 2800 GW d) 3800 GW
8. The offshore sites of wind energy supplies ...... generation:
   a) 3% b) 5% c) 7% d) 10%
9. More than half of the global wind energy capacity is expected to be contributed by 2030 from the...... region:
   a) Africa b) America c) Asia-Pacific d) Europe.
10. A major economic benefit of wind energy is:
    a) Higher cost than fossil fuels
    b) Declining per-unit price
    c) Dependent on fossil fuels
    d) Requires substantial amounts of water
11. Is NOT an environmental benefit of wind energy?
    a) No greenhouse gas emissions
    b) Free of cost
    c) Requires substantial amounts of water
    d) Rural Jobs
12. Wind energy helps to reduce dependency on which energy source?
    a) Nuclear fusion b) Hydropower
    c) Fossil fuels d) Fission
13. Wind energy is often located in .... areas to promote employment:
    a) Urban areas b) Rural and remote areas
    c)Metropolitan cities d) Industrial areas
14. A major challenge associated with wind energy is:
    a) It produces greenhouse gases
    b) It is awfully expensive to install

c) It is intermittent and unpredictable
d) It requires high water use

15. The wind deflects right in the Northern Hemisphere due to:
a) Magnetic effect b) Coriolis effect
c) Friction effect d) Gravity effect.

16. The speed at which a wind turbine starts generating electricity:
a) 0-2 m/sec b) 3-4 m/sec c) 6-9 m/sec d) 10-11 m/sec

17. The speed at which the turbine stops to protect itself from damage.
a) ~5 m/s b) ~10m/s c) ~25m/s d) ~50m/s

18. In a wind turbine, energy conversion takes place from mechanical to electrical using
a) Yaw system b) Gearbox
c) Electrical generator d) Blade

19. The Betz limit of wind turbines to extract wind energy is:
a) 45.3% b) 39.3% c) 59.3% d) 69.3.0%

20. The wind turbine is more suitable for urban areas with variable winds:
a) Horizontal Axis Wind Turbine
b) Vertical Axis Wind Turbine
c) Desert Wind Turbine
d) Offshore Wind Turbine

21. ..... is NOT a category of wind energy systems.
a) Onshore wind energy
b) Offshore wind energy
c) Waterfall wind energy
d) Hybrid wind energy

22. A key benefit of onshore wind energy systems is:
a) No noise pollution
b) Easier and less costly installation and transportation
c) Requires no land area.
d) Operates only in coastal areas.

23. One limitation of onshore wind energy is:
a) They generate constant power
b) They require large land areas
c) No effect on the local ecosystems
d) Always cheaper than offshore systems

24. Offshore wind turbines are usually installed:
    a) In deserts
    b) In large water bodies like oceans and lakes
    c) On mountain peaks
    d) Inside forests
25. Floating wind turbines differ from traditional offshore turbines because:
    a) They are anchored to the seabed only in shallow water
    b) They use mooring lines to stay in place in deeper water
    c) They are smaller in capacity
    d) They cannot be used in the deep sea
26. The capacity factor of offshore wind farms compared to onshore farms is generally:
    a) Zero b) About the same
    c) Higher (up to 50%) d) Lower
27. .... is an environmental concern related to offshore wind farms:
    a) Noise disturbance to marine mammals
    b) Increased fossil fuel use
    c) Air pollution
    d) Visual pollution in cities
28. Wind-solar hybrid systems are beneficial because:
    a) Both generate energy only during the day
    b) Wind is stronger at night and solar during the day, complementing each other
    c) Solar power replaces wind power fully
    d) They need separate infrastructures
29. In wind-battery hybrid systems, the purpose of batteries is:
    a) To reduce noise pollution
    b) To store excess wind power for use when wind is low
    c) To power the turbines directly
    d) To give rest to wind turbines for a longer life.
30. A major challenge of hybrid wind energy systems is:
    a) Less power generation
    b) Short lifespan
    c) Complex control strategies and high investment
    d) No integration with the grid needed

31. The payback period of wind turbines is approximately:
    a) 5-6 months b) 2-3 years
    c) 5-6 years d) 20-25 years
32. A major environmental impact of wind turbines on wildlife is:
    a) Adverse look
    b) Collision risks to birds and bats
    c) Soil erosion
    d) Water pollution
33. Bat collisions with turbines can be reduced by:
    a) Using bright lights at night
    b) Emitting high-frequency sounds to deter bats
    c) Decreasing turbine height
    d) Putting up notice boards to stay away.
34. The noise generated by wind turbine blades is:
    a) Mechanical noise only
    b) Aerodynamic noise (whooshing sound)
    c) No noise
    d) Chirping sound
35. The increase in efficiency and reduction in weight in modern turbine blades are due to
    a) Steel towers only
    b) Plastic mixed with steel
    c) Carbon fibre and composite materials
    d) Bamboo mixed steel blades.
36. The major advantage of vertical-axis wind turbines (VAWTs) is:
    a) Require a large space
    b) Cannot be used in urban areas
    c) Components near the ground for easy maintenance
    d) Less efficient in turbulent winds
37. Floating wind turbines are mainly used because:
    a) Noise pollution
    b) Most offshore wind resources are in deep water (>60m)
    c) They are cheaper than onshore turbines
    d) Easy maintenance
38. A key benefit of smart grid integration with wind energy is:
    a) Ignoring weather conditions

b) Real-time power supply-demand balancing
c) No storage needed
d) Easy maintenance

39. AI in wind energy helps by:
a) Increasing downtime
b) Predictive maintenance and weather forecasting
c) Decreasing turbine noise
d) Decreasing turbine lifespan

40. The world's largest floating wind farm is located in:
a) India b) USA c) Norway d) China.

41. The wind is air movement due to:
A. Ice B. Rain C. Heat difference D. All of the above

42. .... of the world's electricity demand is fulfilled by wind energy.
A. 2% B. 8% C. 18% D. 28%

43. .... country has the largest installed capacity of wind power.
A. USA B. China C. Norway D. India

44. China expects to install .... of wind power by the end of this decade.
A. 250 GW B. 350 GW C. 450 GW D. 550 GW

45. Indian states are leading in wind energy development.
A. Punjab and Bihar B. Tamil Nadu and Gujarat
C. Assam and Rajasthan D. Kerala and Goa

46. To motivate investors, the.......... policy assures a fixed price per unit of electricity is assured.
A. Net Metering B. Carbon Credit Policy
C. Feed-in-Tariffs D. Renewable Energy Certificates

47. The international agreement targets net-zero carbon emissions by 2050.
A. Kyoto Protocol B. Paris Agreement
C. EU REPower Plan D. COP28

48. The Hornsea Windfarm has installed a wind power capacity.
A. 1.6 GW B. 2.6 GW C. 3.6 GW D. 4.6 GW

49. ....of Indian is the world's largest hybrid renewable energy park project.
    A. Jaisalmer Wind Park
    B. Ladakh Solar Park
    C. Gujarat Wind-Solar Hybrid Park
    D. Tamil Nadu Wind Corridor
50. The global investment in the wind sector in 2023, as per the IEA, was:
    A. Rs 1,400 billion
    B. Rs 10,400 billion
    C. Rs 14,400 billion
    D. Rs 25,400 billion

Answer Keys:

1. B, 2. C, 3. A, 4. A, 5. B, 6. C, 7. B, 8. C, 9. C, 10. B, 11. C, 12. C, 13. B, 14. C, 15. B, 16. B, 17. C, 18. C, 19. C, 20. B, 21. D, 22. B, 23. B, 24. B, 25. B, 26. C, 27. A, 28. B, 29. B, 30. C, 31. C, 32. B, 33. B, 34. B, 35. C, 36. C, 37. B, 38. B, 39. B, 40. C, 41. B, 42. B, 43. B, 44. C, 45. B, 46. C, 47. B, 48. C, 49. C, 50. C

आपो हि ष्ठा मयोभुवस्ता न ऊ॒र्जे द॑धातन महे रणाय चक्षसे।

यो वः शिवतमो रसस्तस्य भाजयते हनः। उशतीरिव मातरः।

Rigveda (10.9.1)

Meaning:

Water brings us joy and bliss. May water protect us and sustain us forever. May the purifying essence of water be shared with us like a loving Mother who nourishes children.

Chapter 5

HYDROPOWER ENERGY

Learning Outcomes:

After reading this chapter, the learner will be able to:

1. Explain the principle of hydropower generation.
2. Identify the key component of the hydropower plant.
3. Compare distinct types of hydropower plants in terms of power output and site selection.
4. Analyse the social, economic, and environmental impacts of large hydropower plants.
5. Assess the role of a Pumped Hydro Storage plant in meeting the peak load.

5.1 Introduction

Thousands of years ago, human civilizations had evolved at the banks of rivers. Water has always been the lifeline of any society. Energy production with the help of water has played a vital role in human history. The most prominent way of harnessing energy from water is to convert the potential energy of stored water at height into kinetic energy. The kinetic energy of flowing water is then converted to electricity with the help of turbines and electrical generators. Hydropower has reduced our dependence on fossil fuels, offering other advantages like energy storage, grid reliability, flood management, and water management.

Real Inspiration

Ralegan Siddhi, a drought-prone village, was revived by local women under Anna Hazare's movement through a local Self-Help Group (SHG). Women gained financial independence through improved crop yields. Ralegan is now considered one of the ideal villages. A true Gram Swaraj of Gandhian philosophy

Source: NABARD Rural Stories, 2018

Basic Principle:

The working principle of hydro energy is quite simple. The rainwater or river water is stored behind a big dam. These dams are mostly situated in the mountainous regions to have natural walls to store water at a higher altitude. The stored water from the dam flows from pipes called penstocks to rotate the turbines, which convert the kinetic energy of flowing water into mechanical energy. Rotating turbines are connected with electrical generators, which convert the mechanical energy into electrical energy.

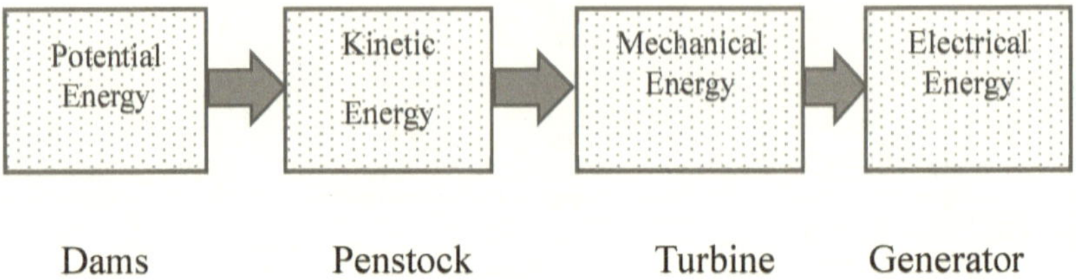

Figure 5.1 Energy Conversion in a Hydropower Plant

Historical Evolution: Even thousands of years ago, our ancestors were harnessing the energy associated with water in several ways. Just like Windmills, water wheels were also used to grind the grains, saw wood, and perform other mechanical work. People in Europe were using water wheels for milling, but all these tasks were primitive and were used to reduce labour.

The modern application of hydropower plants for the production of electricity began in the 19th century. The world's first hydropower plant started to work in 1882, on the Fox River in the USA. It was used to generate sufficient electricity to supply two paper mills. By the beginning of the 20th century, people had started to understand the importance of free electricity generated with the help of hydropower plants, along with other benefits like flood management, irrigation, and providing water in drought and water-scarce areas. The Hoover Dam and Grand Coulee Dam projects were a few of the eye-catching projects during this era. Post World War II, hydropower projects expanded across the globe to tap the free energy of water. We can understand the importance of hydropower plants from the fact that, in the global energy mix, hydropower is the biggest contributor of renewable energy.

Key Statistics: As said above, hydropower energy is the largest renewable energy source in the global energy mix as of today. According to data available from the websites of the International Renewable Energy Agency (IRENA) and the International Hydropower Association (IHA), global installed hydropower capacity has exceeded ~1360 GW, which is about 16% of the world's electricity and more than ~55% of the world's renewable energy *(as per the IHA 2023 report)*. Major countries exploring hydropower energy are China, Brazil, Canada, the USA, and India. The world's largest hydropower plant is located at the Three Gorges Dam, China, with a capacity of 22.5 GW. The major advantages of hydropower plants are a longer life of typically 50 years, higher conversion efficiency, the ability to work as base load and peak load, and minimum maintenance.

Despite these advantages, hydropower expansion has slowed down due to environmental concerns, heavy construction costs, and a dearth of suitable sites. However, small and micro-hydropower plants are being deployed, which require lesser construction costs, a smaller time frame, and are more suitable for rural applications.

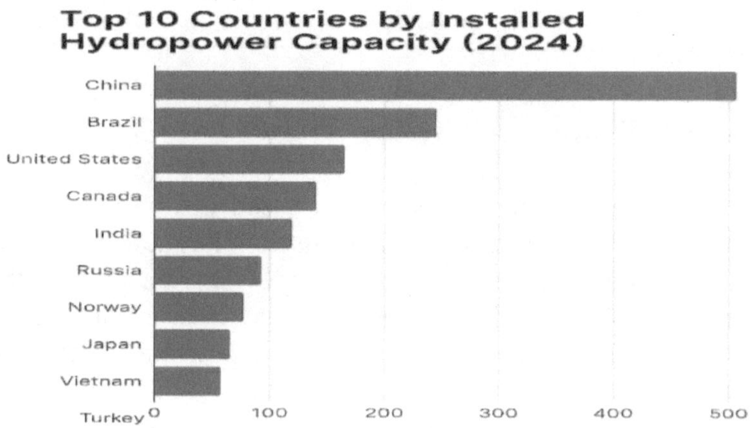

Figure 5.2 Top 10 countries in installed hydropower Plant capacity
(Source: IHA's 2024 World Hydropower Outlook *and REN21's* Renewables 2024 Global Status Report)

Hydropower plants are more reliable in comparison to other renewable sources like solar and wind, which are intermittent in nature and unpredictable. Hydropower plants can supply electricity consistently and are better suited to supply even peak loads due to the ability to control water flow. Pumped storage hydropower systems are the most suitable option to balance the grid during peak demand, making hydropower plants a vital asset over other renewable resources.

Impact: They also directly contribute to the United Nations' Sustainable Development goals: Clean water and sanitation (SDG 6), Affordable and clean energy (SDG 7), Industry, innovation, and infrastructure (SDG 9), and Climate action (SDG 13). They are often multipurpose to supply drinking water, manage floods, and provide water for irrigation. Thus, they gave a multi-dimensional social, economic, and ecological impact.

Hydropower plants have a unique role in combating climate change. As the dependency on fossil fuels is reduced, hydropower plants help to reduce global warming by reducing greenhouse gas emissions. It can also generate jobs in this sector for the peripheral area residents. It can also boost the agricultural activities in the downstream area. However, it may disturb the ecological system and social life due to the relocation of people in and around the catchment area.

5.2 Types of Hydropower Systems

Hydropower systems can be classified based on the capacity of power generation and the technology used. They are large-scale hydropower plants, small and micro hydropower plants, run-of-the-river, reservoir-based, and pumped storage hydropower plants. They are installed depending on suitability, potential of the site, water availability, and financial and environmental considerations. The following is the classification based on the capacity of the power plant. However, it may differ for various countries.

Sr No.	Category	Capacity
1	Large Hydropower plant	Above 100 MW
2	Medium Hydropower plant	25 MW -100 MW
3	Small Hydropower plant	1 MW – 25 MW
4	Mini Hydropower plant	100kW - 1 MW
5	Micro Hydropower plant	Up to 100 KW
6	Pico Hydropower plant	Up to 5 kW – 10 kW

Table 5.1 Typical hydropower Plant capacity [14]

Now, let us explore a few of these plants in detail to get information at a glance.

5.2.1 Large-Scale Hydropower Plants

Large-scale hydropower plants are generally referred to as plants with a capacity of more than 100 MW. These are usually suited to perennial big rivers with continuous water flow. These require huge finances for constructing large dams and reservoirs. Since these are big in size, production, and area, they can supply power and water to a large peripheral area. The power output of these plants is generally connected to the national grid. The world's largest hydropower plants are the Three Gorges Dam of China (22,500 MW), the Itaipu Dam of Brazil and Paraguay (14,000 MW), and the Grand Coulee Dam of USA (6809 MW) *(as per USBR data).*

Disadvantages:

1. These plants are of low capacity, making them unsuitable for the large industrial sector or urban electrical requirements.

2. Since these plants are often located on small rivers, they may face variable water flow, and dry seasons may have limited power supply.

Applications:

1. These small plants help reduce dependency on diesel generator sets.
2. These are helpful in the electrification of energy-deprived rural areas.
3. These plants can fulfil the electricity needs of a small industry.
4. These plants are ideally suited for remote resorts or off-grid areas

> ***Activity**: Observe the area around your village. Can you spot any sites which may have a potential to produce power to at least power one home or farm? Calculate possible expenditure and payback period.*

Run-of-the-river and Reservoir-based systems:

Hydropower plants can also be classified based on how they manage water flow. There are primarily two types of systems that exist, viz, Run-of-the-river (RoR)and reservoir-based systems.

Run-of-the-river (RoR) System:

In this case, no reservoir or dam is required to be constructed, but part of the river water is channelled through penstocks to rotate the turbine and generator. This water is then channelised to the river downstream. These plants can not utilize the full potential of the river water. They are small in capacity. They are site-specific and are suitable for areas with steep slopes, like in the Himalaya and the Andes mountains. They are environmentally friendly.

Advantages:

1. These plants require incredibly low capital investment.
2. They have a minimum carbon footprint.
3. They do not require relocation or people.
4. They do not require a long construction time.

Disadvantages:

1. These plants have limited facilities to regulate the water flow, hence can not be used for balancing peak load.
2. These are frequently affected by water output in rivers.

Reservoir-based systems:

These hydropower plants require the construction of a large dam to store water. This helps to regulate the water flow and thus makes these systems ideal for base load as well as to supply the peak load. These systems are often large and are for multipurpose use, like irrigation, drinking water, and managing floods.

Advantages:

1. These types of hydropower plants can provide reliable power output in all seasons.
2. It can address the load demand by regulating water flow.
3. Since they are multipurpose, they are a boon to supply drinking water in water-scarce areas. For example, the Sardar Sarovar Narmada Dam, situated in the western part of India, supplies drinking water to the people living in the desert areas of Kutch, Gujarat.

Disadvantages:

1. Since these plants are big, they require a big capital investment for the construction of the dam, purchase of turbines, generators, penstocks, and other accessories.
2. Sedimentation over a prolonged period may reduce the storage capacity.
3. Since these are spread across a wide area, they may require the relocation of the local people.
4. A limited number of suitable sites are available.

5.2.2 Pumped Storage Hydropower

Pumped storage hydropower is a special type of system in which stored water is used as a *giant battery* to meet the peak load in emergencies. During the low electricity demand, excess electricity from other renewable energy sources like solar and wind is used to pump the water from the lower reservoir to the upper reservoir. During peak demand, this pumped water is now released back to the lower reservoir through the generator to generate electricity. Figure 5.3 is represented in a flow format for easy understanding.

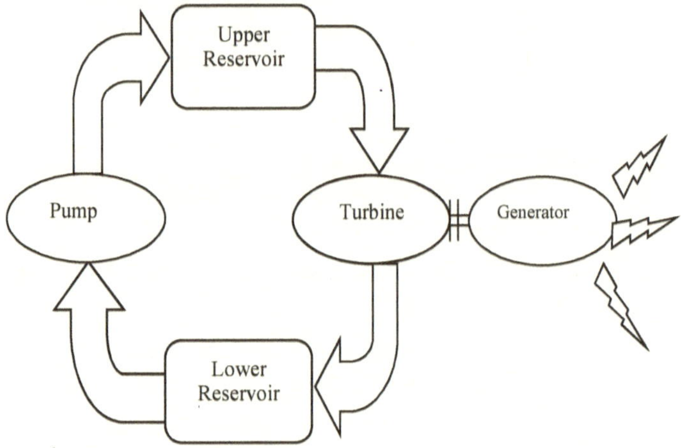

Figure 5.3 Block Diagram to Show Water Flow in PSH Plant.

This system does not consume water, but it merely acts to stabilise the grid by using excess energy during low demand at times of high demand. Thus, we say, here, water acts as a giant battery! A few leading examples are Bath County PSH from the USA (3,003 MW, the largest PSH in the world), and Tamahara Dam, Japan [35].

Advantages:

1. This system is highly effective in meeting the peak demand rapidly.
2. This system has a long life span of about 70 to 100 years.
3. This system has 70 % to 85 % efficiency.

4. This is a quite simple system, having no complex mechanism.
5. This system can integrate other renewable energy sources like solar and wind to reduce the carbon footprint.
6. It can also provide a power supply during outages.
7. This system can also be helpful in voltage and frequency regulations in smart grids.

Disadvantages:
1. Pumped storage hydropower plants require high initial investments and long payback periods.
2. This system can be deployed only at specific locations with suitable topography and water availability.
3. There may be environmental constraints.

The following table 5.2 is a comparative analysis between several types of hydropower plants:

Sr	Type of Hydropower plant	Scale	Storage	Main Benefits	Main Constraints
1	Large Scale	>100 MW	Yes	High Capacity	Environmental
2	Medium Scale	25 MW-100 MW	Yes	Medium capacity	Environmental
3	Small/Micro	<25 MW	Limited	Local energy access	Limited capacity
4	Run-of-the-river	Variable	Minimum	Eco-friendly	Seasonal
5	Reservoir based	Any	Yes	Regulated Output	Excessive cost
6	Pumped Hydro	Variable	Yes	Load balancing	High initial cost

Table 5.2 A Comparative analysis of Hydropower plants

5.2.3 Components of Hydropower Plants

The basic working principle of all hydropower plants is to convert the potential energy at high altitude into kinetic energy and then utilize it as mechanical energy to rotate the turbine-generator set to produce electrical energy. Irrespective of size, type, and configuration, the basic working principle and components of hydropower plants remain almost the same. Let us now explore these components in detail:

- Dam: This term is specifically used for the special concrete structure. It is built across the river to store water. By increasing the dam height, more water will be stored behind, enhancing the potential energy of water, which is particularly important for the generation of electricity. Basic functions of dams are to create head height for energy conversion, to store water, and also to control floods, providing irrigation facilities and navigation.

  Mainly, there are three types of dams. Gravity dams use their weight to block water. Bhakhra Dam is one such dam. Arch dams are curved in nature, which transfers the water pressure to the canyon walls. Idukki Dam is an example of an Arch dam. Buttress dams use supports in the downstream to withstand water pressure.

- Reservoir: It is the artificial lake created behind the wall of a dam. Water is stored in the reservoir. Sedimentation occurs in the reservoir, where heavy particles settle at the bottom to increase the life of the turbines. Sometimes reservoirs are large, leading to environmental challenges, submergence of land, methane emissions from organic decay, and may cause disturbances in the aquatic ecosystem.
- Intake Structure: It refers to the mechanism that directs the water to flow into the penstocks. Typically, they are fitted with screens to filter debris, Control gates for the regulation of water flow, and a few flow meters and sensors for monitoring purposes.

- Penstocks: These are the large pipes that transport water from the reservoir to the turbines. They may be made up of concrete, steel, or fibreglass, depending on the pressure of water. Penstocks should be able to withstand high pressure and the velocity of water. They are equipped with valves to tackle emergencies and for maintenance. Penstocks should be designed in such a way as to minimise the friction and turbulence. The diameter, length, and slope affect the power generation of the hydropower plant.
- Turbines: Turbines are the key components of any hydropower plant, where the kinetic energy of water is converted into rotational mechanical energy. There are two types of turbines. Impulse turbines are used for high-head and low-flow sites. The Pelton wheel is an example of an impulse turbine. Reaction turbines are more efficient in low and medium-head sites. Francis's turbine and Kaplan turbine are examples of reaction-type turbines. Water from the penstock strikes the turbine blades, which tend to spin, leading to the rotation of the generator shaft.
- Electrical Generator: A Generator converts mechanical energy provided by a turbine into electrical energy using the principle of electromagnetic induction. Generators consist of a rotor, a stator, an exciter, and other accessories. The relative motion between the copper windings of the stator and rotor magnetic fields induces alternating current in the stator. This induced power is then transmitted through a network of transmission lines to consumption points. Generators may be Synchronous or Asynchronous. A synchronous generator operates at a constant speed and is more common in modern grids to maintain a constant supply frequency. Asynchronous generators are used in isolated systems. Generators typically have high efficiencies exceeding 90%; they may be equipped with a cooling system and damper system, depending on their size.
- Control Systems: Most of the modern generators are equipped with automated control systems to control the turbine operation (for: flow, speed, blade pitch) to enhance the generator performance. Control systems monitor

variation in voltage and frequency, and water level in the reservoir. It also remotely supervises and diagnoses the faults. Supervisory Control and Data Acquisition (SCADA) type systems also help in remote management to optimise performance. Now, smart sensors and IoT-based predictive tools are also used to integrate cybersecurity.

- Grid Integration: Generated electricity should be transmitted to load centres for utilisation. The power grid is used to transmit the generated electricity. It requires step-up transformers to raise the voltage level to reduce losses over long distances. Transmission and distribution lines are used to transmit the power. Step-down transformers are used to reduce the voltage level to make it compatible with consumption points.

> ***Thinking Corner***: *Suppose your home, along with your village and farm, are being submerged in a proposed dam project. The government is offering suitable compensation. What would be your opinion?*

5.3 Advantages of Hydropower

Hydropower is one of the oldest forms of renewable energy. Hydropower addresses the challenge of carbon emissions while providing energy security, climate adaptation, and socio-economic development. Hydropower helps in energy sustainability, reduces environmental impact, and improves water resource management. The following are the key benefits of hydropower plants:

- Renewable and reliable source of energy.
- Inexhaustible energy source.
- High energy conversion efficiency (Hydro >90% , Thermal ~ 30-40 %)
- Consistent and Predictable output.
- Longer life up to 50 to 100 years.
- Low operating costs.
- Flexibility (from a few kW in small run-of-the-river systems to GW-scale dams)

- Carbon-neutral electricity generation. (Coal~ 820 g, Solar PV ~ 45 g, Hydropower ~24 g per equivalent kWh).
- Carbon offset and renewable credits.
- Fast ramp-up option in load balancing.
- Reliable source as spinning reserve in case other sources fail.
- Black start capacity to restart the grid after a major outage.
- Plays a key role in integrating other intermittent renewable energy sources like wind and solar during cloudy or night hours.
- Much cheaper option to meet peak load through pumped storage power (PSH).
- To regulate the water flow of rivers to mitigate flood risks.
- Boosts agricultural productivity in downstream areas. Other areas may benefit from canal networks. It also supports drought-prone areas.
- Drinking water facility for millions of people.
- Industrial water supply to boost production in industries that require a steady water supply, like textiles and paper plants.
- Inland water navigation for transportation of goods.
- The reservoir and surrounding area may be developed as a tourist site to promote local tourism and fishing to enhance the local economy.

5.4 Challenges and Environmental Concerns:

Even when hydropower is often regarded as one of the most reliable and environmentally friendly renewable energy sources, it has a few challenges and concerns to consider before its implementation. The following are the few points required to be considered:

- It alters the river habitat. It blocks the free-flowing river into the stagnant reservoir with reduced downstream flow, leading to reduced biodiversity.

- It becomes a barrier to many river fishes like salmon and eels, which undergo seasonal migrations for reproduction. This may lead to the collapse of fisheries.
- It creates an impact on aquatic and terrestrial food webs.
- Forced relocation of local communities is also involved in constructing a dam. For example, the Sardar Sarovar Dam has led to the displacement of ~40,000 families (~2 lakh people), while China's Three Gorges Dam displaced more than 13 lakh people.
- Many sacred sites, ancient temples, and archaeological remains may be submerged, leading to the loss of historical and cultural sites.
- It may lead to conflicts and resistance movements to oppose the submergence of land and village from local villagers who have resided there for thousands of years.
- Reduced water capacity due to sediment accumulation in the reservoir, leading to reduced power supply.
- A large amount of water may provoke seismic activities and Earthquakes.
- Reduced water storage and power generation during drought or low rainfall in the catchment area.
- Flood-induced damage during extreme rainfall in the catchment area.
- Reduction in glaciers and melting of ice are posing new challenges to maintaining water availability.
- Increased global warming has led to higher water evaporation.
- In case of dam failure, the artificial flood may devastate the urban and rural population in the downstream areas.

> ***Thinking Corner**: What happens if, due to climate change, rainfall reduces, making dams run dry? How resilient is hydropower in the present climate -uncertain future?*

5.5 Technological Innovations in Hydropower

The technological renaissance has equipped hydropower with many modern, innovative tools to meet the demands of sustainability, resilience, and decarbonization. To reduce the impact on biodiversity, fish-friendly turbines are introduced, having a slow rotational speed to reduce the blade strike and wider flow passages to prevent fish from being trapped. An example includes the Alden turbine, which has fish survival rates of more than 98%. With the advent of Industry 4.0, hydropower plants have undergone digital transformation to reduce operational costs and enhance environmental compliance. This includes predictive maintenance, real-time flow management, remote operations and remote control, and cybersecurity. To meet the challenge of the unpredictability of wind and solar, hydropower plants are hybridised with them with the latest technology for better power management and load balancing. Underwater and tidal turbines are also being explored to get the energy from freely flowing water rather than being blocked in a reservoir.

5.6 International Collaboration and Development

Developing countries often receive support from Multilateral Development Banks (MDBs) and International Organizations for project financing, technical assistance, feasibility studies, and environmental and social impact assessment support. Major players are the World Bank, Asian Development Bank (ADB), African Development Bank (AfDB), and International Finance Corporation (IFC).

International frameworks are constantly advocating for sustainability, leading to the Hydropower Sustainability Standard by the International Hydropower Association (IHA). The Paris Agreement has also paved a positive path to boost renewable energy sources like hydropower plants to reduce GHG emissions.

Rivers often flow through more than one country. Building a dam on the river to block water often fuels conflicts between different nations. International organisations help solve these types of conflicts and manage river basins, power trade agreements to improve regional energy security. For example, the Eastern African Power Pool (EAPP) helps member countries with electricity trading using large-scale hydropower plants like Ethiopia's Grand Renaissance Dam. International cooperation also enables knowledge sharing, training the workforce, sharing innovative technologies, and real-time data monitoring to tackle floods and droughts.

5.7 Future of Hydropower in Sustainable Development:

Ever-increasing demand for energy and ever-increasing global warming leading to climate change mandate the compulsory role of renewable energy sources like hydropower plants. Hydropower offers a low-carbon, resilient, and sustainable solution for the future. Now, let us discuss the role of hydropower in sustainable development for tomorrow.

- **Low Carbon profile:** Most energy sources require burning of the fuels to produce energy, while hydropower converts the kinetic energy of water to electrical energy. So, the negligible carbon footprint is associated with it. While discussing the advantages of hydropower, we have already stated that hydropower plants have remarkably high energy conversion efficiency. It has an average emission factor of just 24 g CO_2/kWh in comparison to 820 g CO_2/kWh of coal. It can very easily adjust the supply and demand during peak loads and help in managing frequency and voltage stability. Pumped storage hydropower plants are now being utilized as large water-based batteries! These PSH plants have become essential, affordable parts to increase the flexibility of the grid in countries like India, China, and Australia.
- **Integration with Climate Adaptation:** We are witnessing the impacts of climate change, such as varying rainfall patterns, faster glacier melting rates, and others, which are affecting the water flow. Hydropower can help with flood control, management, and support irrigation, which is essential in changing rain patterns. Hydropower can also be classified as climate-resilient infrastructure due to its long lifespan

and co-benefits. It can also provide green infrastructure synergy through floating solar plants. It also helps to increase the water table of the catchment area and its surroundings.

- **Community-based Micro Hydro Projects:** These small plants may prove a boon for remote areas, where ample amounts of solar radiation are either not available or it is difficult to harness solar energy, like mountainous regions. They are typically under 100 kW in capacity and are useful to meet the needs of homes, schools, hospitals, grain mills, and streetlights etc.

5.8 Case Studies:

In the global energy map, hydropower is playing a significant role, ranging from small micro-hydro systems to gigantic mega dams. They play a significant role in shaping the socio-economic paradigm of society. Let us discuss a few examples in this regard that have created a significant impact on the global stage. **The Three Gorges Dam** is located on the Yangtze River in China. It is considered the world's largest hydropower plant with an installed capacity of 22,500 MW. Its reservoir length is more than 600 km, and it generates more than 95 TWh per year. However, it has displaced more than thirteen lakh people. It offsets more than 30 million tons of CO_2 per year, thus significantly reducing the carbon footprint. It also helps to manage the floods in the downstream of the Yangtze River.

Bhakra Nangal Dam was commissioned in 1963 on the Sutlej River of India. It was one of the largest multipurpose projects in India. It has benefited millions of people in western India for drinking water, irrigation, and agricultural industries, along with fulfilling the need for rural and urban electrification. Its installed capacity is 1,325 MW. Once it was termed as "Temple of Modern India" by the then prime minister of India, Shri Jawaharlal Nehru, but now it is facing concerns over sedimentation and declining water storage capacity.

The Tehri Dam project is located on the Bhagirathi river, with an installed capacity of 1,000 MW. Additionally, it is one of the tallest dams in India, with a height of 260.5 meters. It supplies electricity and drinking water to New Delhi, the capital of India. It also helps to irrigate the farms of Uttar Pradesh. Since it is located near a high-power-consuming area, its role is significant in grid balancing through the energy storage of an additional 1,000 MW pumped storage hydro system. A few controversies are also linked with this project, related to its location in a seismic zone and the submergence of Tehri town.

Micro hydro projects in Nepal have transformed rural livelihoods through electrification and energy access. Nepal has taken the benefit of its mountainous terrain and deployed more than 3,000 micro hydro systems, leading to the electrification of more than 3,00,000 households, reduction in the use of fossil fuel sources like kerosene and wood. It has generated local employment and empowerment of local women in weaving, Agri-processing, and carpentry industries. The best part is that these micro-hydro hydropower plants are based on community ownership models, which ensure direct supervision and immediate maintenance.

Norway's hydropower-driven grid is one of the major case studies to learn a lesson in implementing hydropower at its best. Norway is a global leader in hydropower, which generates more than 90 % of its electricity needs from hydropower plants. It has an installed hydropower capacity of more than 33,000 MW [36]. Norway prominently uses PSH plants to stabilize the grid while meeting the peak demand.

Activity: *Research the Machchu dam failure. Find the number of people who died and the total loss of property. What may be the reasons for a dam failure?*

5.9 Summary

- Hydropower is the largest form of renewable energy sources, contributing ~17% of global electricity generation. (IEA report 2023)
- It is a reliable base-load power plant helpful in stabilizing the grid.
- **PSH** or pumped storage hydropower plants are a ridiculously cheap, reliable, and prompt option to meet the peak demand for electricity.
- The main components of hydropower plants are the reservoir, dams, penstocks to carry water from the dam to the turbines, a sluice gate to control water flow, generators, and a control system.
- Major advantages of hydropower plants: they provide reliable, renewable, and flexible power. It supports peak power with a minimal carbon footprint at a low cost. Its greenhouse gas emissions are negligible in comparison to other traditional sources of energy. It also provides the benefits of drinking water, irrigation, fishing, and flood control.
- Ecosystem disruption, fish migration issues, sedimentation, social displacements, Earthquakes, and changes in river flow patterns are key issues to address while implementing the project of large hydropower.
- Fish-friendly turbines, solar-hydro hybrids, tidal, and underwater turbines are trending innovations for future hydropower plants.
- Kyoto Clean Development Mechanism, Green Bonds, national and international regulations, and frameworks are essential to promote this non-polluting energy.
- The levelized cost of Electricity (LCOE) of hydropower plants is extremely low due to lower operational maintenance costs and longer lifespans.
- Hydropower plants play a significant role in achieving UN SDG 7, which is Affordable and Clean Energy. While inclusive development is promoted through community-based micro hydro plants.

-:EXERCISE:-

Section A: Short Answer Questions

1. Compare run-of-the-river and reservoir-based systems.
2. State the typical capacity of a micro hydropower plant.
3. List two advantages and two limitations of pumped hydropower systems.
4. State True or False: "Small and Micro hydropower plants are suitable for remote regions."
5. List the five largest hydropower plants in the world regarding installed power capacity in GW.
6. List the five largest hydropower plants in India regarding installed power capacity in GW.
7. The hydropower (including all types) generates...... % of global electricity.
8. Approximate energy conversion efficiency of hydropower plants is.......... %.
9. State the role of hydropower plants in flood control.
10. List types of hydropower plants along with approximate capacity.

Section B: Long Answer Questions

1. Explain several types of hydropower plants.
2. Discuss how PSH can help peak load.
3. Explain the advantages of large hydropower plants.
4. State the role of small hydropower plants.
5. Mention the advantages of PSH for climate adaptation.
6. Discuss the demerits of large hydropower plants.
7. Explain innovations in hydropower plant turbines.
8. State the role of international hydropower plants in mitigating climate change.
9. List and explain major components of large hydropower plants.
10. Identify and locate probable future sites of large hydropower plants in India.

Section C: Multiple Choice Q and A

1. Hydropower plants harness .... form of energy.
   A. Nuclear B. Thermal
   C. Potential D. Hidden electrical
2. The .... converts water flow into mechanical energy is:
   A. Boiler B. Turbine
   C. Condenser D. Reactor
3. .... is NOT a type of hydropower plant.
   A. Run-of-river B. Biomass
   C. Pumped storage D. Storage
4. A hydropower plant may be considered as:
   A. Slow starter B. Carbon Sequester
   C. A renewable source D. None of these
5. The largest hydropower plant in the world is:
   A. The Hoover Dam B. The Bhakhara Nangal Dam
   C. Three Gorges Dam D. The Itaipu Dam
6. Which country has the highest installed hydropower capacity?
   A. Venezuela B. China C. Canada D. Brazil
7. In PSH, water is:
   A. Used only once
   B. Released directly to the sea
   C. Pumped back to an upper reservoir
   D. Stored in underground tanks
8. .... is a benefit of hydropower.
   A. Inconsistent energy output
   B. High carbon emissions
   C. Grid stability support
   D. High operational fuel cost.
9. For Low head, high flow, ...turbines are most suitable.
   A. Francis B. Pelton
   C. Kaplan D. None of these
10. The Francis turbines are:
    A. Impulse turbines B. Reaction turbines
    C. Wind turbines D. Gas turbines

11. The large hydropower dams pose a challenge of:
    A. Deforestation B. Fishing
    C. Displacement of local communities
    D. Government Policies
12. Run-of-river projects have minimal:
    A. Life span B. Energy output
    C. Environmental impact D. Water usage

13. The large hydropower has a demerit of:
    A. Water storage B. Less jobs
    C. Environmental disturbance D. Aesthetic
14. Hydropower generation mainly depends on:
    A. River breadth B. Rainfall and elevation
    C. Water speed D. Water Temperature
15. .... component converts mechanical energy into electricity in a hydroelectric plant.
    A. Motor B. Transformer C. Generator D. Penstocks
16. The Bhakra Nangal Dam is on the.......... river.
    A. Saraswati B. Krishna C. Sutlej D. Kaveri
17. The term "head" in hydropower refers to:
    A. Volume of the dam
    B. Height of the water column
    C. Size of the turbine
    D. Distance from the generator.
18. The common generation capacity of small hydropower plants is:
    A. 10 MW B. 100 MW C. 200 MW D. 300 MW
19. Hydropower contributes significantly to ....global energy goal.
    A. SDG 4 B. SDG 5 C. SDG 7 D. SDG 8
20. Hydropower plants are highly efficient because of:
    A. Low transmission losses
    B. 90 % + conversion efficiency
    C. Free water
    D. Manual operation
21. .... is used to regulate water flow into the turbines.
    A. Surge tank B. Gate valve C. Penstock D. Sluice gate.
22. A major risk during dam failure is:
    A. Water scarcity B. Fishing

C. Flash floods D. Corrosion

23. .... type of hydropower plant is best suited for grid load balancing.
    A. Wave B. Hybrid
    C. Pumped-storage D. Tidal
24. Hydropower is well known for its:
    A. Low operating cost B. Low investment
    C. Water level D. Jobs
25. .... The state of India has high hydropower potential:
    A. Gujarat B. Meghalaya
    C. Himachal Pradesh D. Punjab
26. The sedimentation creates a problem of:
    A. Less fishing B. Better water quality
    C. Reduced dam life D. Increased storage capacity
27. The tailrace is:
    A. A tail part of a turbine blade
    B. A power line end
    C. Channel that discharges water
    D. Water pipe terminal
28. The first hydroelectric project in India was in:
    A. Tehri B. Sivasamudram
    C. Bhakra D. Koyna
29. .... is a micro-hydro application.
    A. 50 kW village supply B. 1 kW Charger
    C. Urban backup system D. 50 MW Mall power.
30. Hydropower is ideal for:
    A. Emergency backup only
    B. Baseload and peak-load supply
    C. Intermittent power only
    D. Battery backup supply
31. The .... energy is stored in the water of the dam.
32. ________ turbines are suitable for low-head, high-flow conditions.
33. The Three Gorges Dam is located in ________.
34. A ________-storage system pump back water in dam.
35. The ________ turbine is used in medium-head applications.
36. Run-of-river projects require ________ reservoirs.

37. The ________ component regulates water pressure in penstocks.
38. The power generation decreases when ________ accumulate in reservoirs.
39. Hydropower plants have high energy ________ (above 90%).
40. An Indian state with significant small hydro potential is: ________.

True/False:

41. Hydropower is a renewable energy source.
42. Francis turbine is used only in nuclear power plants.
43. Penstocks are water-carrying pipes.
44. PSH helps stabilize power grids.
45. All hydropower plants require large dams.
46. A dam is a structure to hold enormous amounts of water in a reservoir.
47. Hydropower plants do not require rainfall.
48. Hydropower requires fossil fuel to start.
49. Large hydropower does not have an environmental impact.
50. Large hydropower plants require a large investment.

Answer Keys:

1. C, 2. B, 3. D, 4. C, 5. C, 6. B, 7. C, 8. C, 9. C, 10. B, 11. C, 12. C, 13. C, 14. B, 15. C, 16. C, 17. B, 18. A, 19. C, 20. B, 21. C, 22. C, 23. C, 24. A, 25. C, 26. C, 27. C, 28. B, 29. A, 30. B, 31. Potential, 32. Kaplan, 33. China, 34. pumped, 35. Francis, 36. Minimal, 37. Surge tank 38. Sedimentation,39. efficiency, 40. Himachal Pradesh *(Uttarakhand has good potential, but as of 2025, Himachal has harnessed it well)*

41. True 42. False 43. True 44.True 45.False
46. True 47. False 48. True 49.False 50. True

"अस्तु नित्यं शक्तेः अक्षयः स्रोतः।"

"Let there be an eternal, inexhaustible source of energy."

(A modern Sanskrit phrase inspired by the Bhagavata Gita)

Chapter 6

OTHER SOURCES OF RENEWABLE ENERGY

Learning Outcomes:

After reading this chapter, the learner will be able to:

1. Explain the working principles of Geothermal, Tidal, Wave, Biomass, and Hydrogen energy.
2. Compare the merits, demerits, and challenges of geothermal, tidal, wave, and hydrogen energies.
3. Discuss the innovations in energy storage and their future possibilities to suit sustainability.
4. Understand nuclear fission and fusion reactions in terms of their role in sustainability.
5. Assess the relevance of these energy systems in meeting climate change.

In previous chapters, we have discussed wind energy, solar energy, and hydro energy in detail. In this chapter, we shall focus on other renewable sources like geothermal energy, tidal energy, wave energy, and hydrogen energy, as well as the current trends in energy storage systems across the globe.

6.1 Geothermal Energy: An inexhaustible Precious Earth Resource

Geothermal energy is the heat acquired from the Earth. This heat originates from the Earth due to many reasons, like radioactive decay, geological processes, movement of plates, heat of the inner core of the Earth, movement of Magma, etc. These sources of energy are getting more attention in recent years due to the increased inclination towards sustainable development and decarbonization goals. Geothermal energy is more preferable than fossil fuels due to its minimal greenhouse gas emissions.

Geothermal energy can be found in hydrothermal reservoirs, geopressured zones, hot dry rocks, and certain magma-based systems. The captured/stored heat deep inside the Earth's crust is taken out by drilling a hole deep into the ground. The heat acquired from the Earth's crust is used for electricity generation if its temperature is more than 150°C, for greenhouse heating, room heating in colder countries, and industrial use if its temperature is about 100°C. As of 2024, the global installed capacity of geothermal power is about 16873 MW [37]. One interesting fact, taking advantage of naturally available resources at its best, Iceland uses geothermal energy to warm more than 90% of its homes! The following pie chart sheds light on the installed capacity of leading countries in this sector. In India, ~340 sites have been identified for the possible geothermal energy extraction by GSI; however, as of this date, no commercial plant exists due to the lower temperature and pressure at the sites. These are located in Himachal Pradesh, Jharkhand, Chhattisgarh, and Gujarat. Research is being conducted under MNRE, India. The following is the classification of geothermal power plants

Type	Principle	Temperature	Characteristics	Example
Dry Steam Power Plants	Use direct steam from a hot reservoir	150°C - 300°C	Located in regions of high geothermal activity.	Lardello, Italy
Flash Steam Plants	Hot water allowed to vaporize into steam at the surface	high temperature >180°C and high pressure	Most common	Cerro Prieto geothermal power plants, Mexico
Binary Cycle Plants	Fluid of a lower boiling point is passed through a heat exchanger using the heat of a low temperature	100 - 170°C	Suited for lower temperature sites	Neal Hot Springs, Oregon, USA

Table 6.1 Classification of Geothermal Power Plants

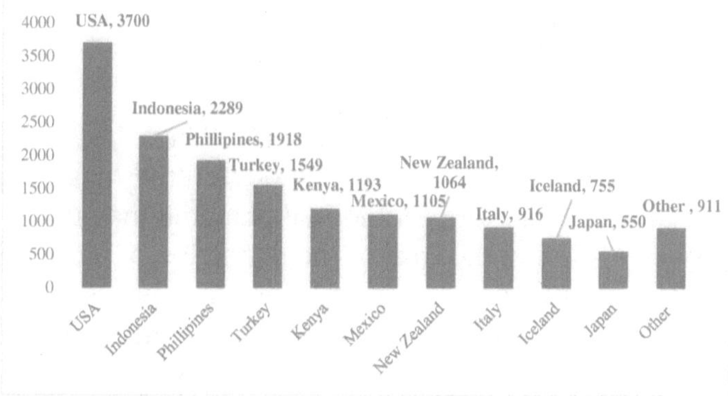

Figure 6.1 Country-wise Installed Capacity of Geothermal Energy

Advantages:

- Geothermal plants emit a small amount of greenhouse gases in comparison to fossil fuels. Binary systems further reduce GHG emissions. So, they are environmentally friendly.
- Geothermal plants require a small amount of land in comparison to solar and wind, so there is no displacement of locals or disruptions to the environment. The sites of geothermal plants are mostly far from residential areas and are not suitable for residential purposes. So, we may say, geothermal plants have a minimal land footprint.
- Geothermal plants are ideal for water-scarce arid regions, as they consume a small amount of water.
- Geothermal plants can provide energy 24x7 continuously without interruptions. So, these are more reliable as base load power plants.

Challenges:

- Geothermal plants cannot be installed at any location, as they are site-specific depending on the Earth's crust characteristics of that location.
- Exploration and drilling wells are an expensive process, which increases the initial capital cost. One cannot be sure about the commercial viability of the identified location.
- Extraction of geothermal energy can trigger minor Earthquakes, so meticulous assessment is needed.

- Ground sinking may occur in the case of long extraction of geothermal fluid without reinjection.

Technological Innovations: Enhanced Geothermal System (EGS) uses the principle of injecting water into fractured rocks to extract heat. This method has more geothermal potential than the naturally occurring reservoirs. In the case of the extraction of oil, it has been seen that hot water is also extracted. This hot water is used for the production of geothermal energy. This also helps in resource circulation. In the case of hybrid systems, a geothermal system is combined with other renewable resources like solar or biomass for efficiency enhancement and load balancing. This method is particularly more useful in the case of off-grid or micro-grid setups. It has been seen that many geothermal reservoirs are rich in lithium. This precious and rare metal can be efficiently extracted from geothermal brine to promote EVs and clean energy.

Role in Sustainable Development: The Geothermal energy sector directly supports many UN Sustainable Development Goals (UN SDGs).

SDG	Detail	Relevance
SDG 7	Affordable and Clean Energy	Geothermal energy provides a renewable, low-cost electric supply along with heating, reduces fuel imports, and assures energy security.
SDG 9	Industry, Innovation, and Infrastructure	Geothermal energy directly promotes sustainable energy infrastructure
SDG 11	Sustainable Cities and Communities	Geothermal energy improves and eases urban life and communities by providing cooling and reducing air pollution.
SDG 13	Climate Action	Geothermal energy helps in the reduction of greenhouse gas emissions, promotes resilience, and local energy independence.

Table 6.2 SDGs and their Relevance Related to Geothermal Energy

Case Studies: Iceland has established its nation as the global leader in the field of geothermal energy by achieving the benchmark of meeting the heating needs of more than 90% of homes through geothermal energy. It is one of the finest examples to showcase urban sustainability through geothermal energy. On the other hand, Kenya is powering more than 40% of its national grid through an 800 MW Olkaria Geothermal Power plant. This has reduced dependency on fossil fuels significantly. India has very few potential sites for the commercial extraction of geothermal energy; however, a demonstration plant of 1 MW has been established to meet the heating needs of the cold desert and energy security in Puga Valley of Ladakh.

> ***Thinking Corner**: If geothermal power plants start extracting geothermal heat on a mass scale, is it possible that the Earth will become cool in the future?*

6.2 Biomass

Biomass is one of the oldest forms of energy known to humans. It refers to the energy that is derived from the basic naturally occurring raw materials like wood, domestic solid wastes, plants and animal wastes, and other organic matters. For centuries, it has played a vital role in the development of human civilization to provide warmth and cook raw food. Since it is a renewable resource, biomass supports the goals of sustainable green energy and helps reduce dependence on fossil fuels if used responsibly. The following (Figure 6.2) are the sources of biomass:

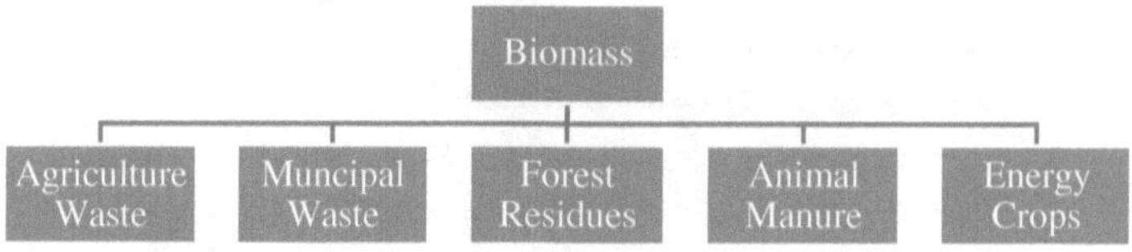

Figure 6.2 Sources of Biomass

- Agriculture residues like rice husks, corn stover, wheat straw, and sugarcane bagasse. These are available in enormous quantities almost every year after the end of the crop season.

- Forest residues include sawdust, bark, woodchips, and wood residues. These are generally by-products of industries related to plywood and wood. You might have also seen a carpenter cutting extra wood from the sheet of plywood to prepare your door!
- In rural areas, people domesticate animals for milk and farming. Their dung is used as manure. This animal manure can be a reliable source of biogas if treated with anaerobic digestion. People in India would have seen biogas plants in almost every village. These plants were promoted by the Government of India for the energy security of the rural population.
- With the increasing population in cities, municipal waste is also increasing. The urban wastes, like food scraps, paper, and other organic waste, can be processed through an incinerator or gasification to get energy. Many city administrations are exploring the options to generate fuel from this urban waste to run city buses.
- Governments across the globe are also exploring the options of dedicated energy crops like miscanthus, switchgrass, and jatropha plantation in arid or marginal lands to acquire high-yield biomass.

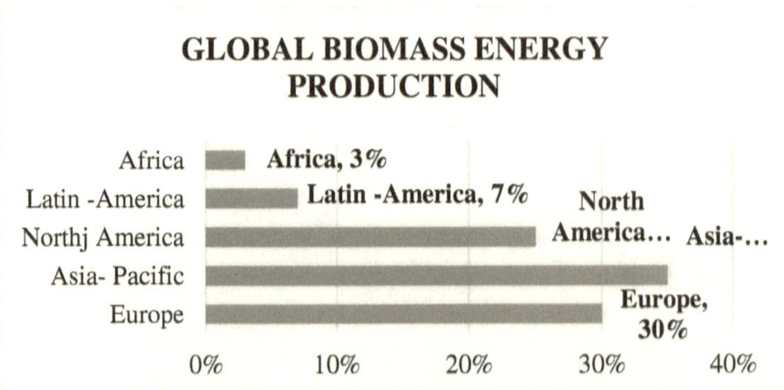

Figure 6.3 Trend of Biogas Energy Production in Various Parts of the World

Biomass Conversion Technologies: There are various technologies to convert biomass into usable energy, like heat, fuel, and electricity. Let us discuss a few of them:

1. Direct Combustion: This is the most common method to harness energy from biomass. In this case, biomass is directly burned to produce heat, which is then used to produce steam to drive a turbine to generate electricity. In this case, biomass is used in place of coal to generate steam in boilers.
2. Thermochemical Conversion: In this case, biomass is decomposed in low or no oxygen to produce bio-oil, biochar, or syngas. Pyrolysis or gasification is generally used for thermal decomposition.
3. Biochemical Conversion: In this case, biomass is broken down with the help of processes like fermentation (to produce ethanol) or anaerobic digestion (to produce biogas) with the help of microorganisms.
4. Transesterification: In this method, vegetable oils or animal fats are treated with chemical processes to produce fatty acid methyl esters (FAME), which can be used as diesel. Since this diesel is prepared from organic materials (bio), it is also popularly known as biodiesel [14].

Conversion method	Efficiency	End product	Emissions	Capital Cost
Direct Combustion	20-40%	Heat, Electricity	Medium	High
Anaerobic Digestion	30-40%	Biogas	Medium	High
Gasification (for combined heat/power)	35-50%	Syngas, Electricity	Low	Medium
Transesterification	>90%	Biodiesel	Low	Low

Table .6.3 Comparison of various conversion methods of biogas

Advantages of biomass in Green Technology and Sustainable Development

We may consider biomass as carbon-neutral because the CO_2 produced during the combustion of biomass is offset by the plants during their growth. This does not happen in the case of fossil fuels. Extraction of energy from biomass also helps in the reduction of wastes from urban as well as rural areas because agricultural and municipal wastes are treated and decomposed; otherwise, this waste would have contributed to pollution and methane emissions. Small-scale plants can be established with the help of biomass to promote rural electrification and rural development. Biomass also helps to get cooking gas through animal and domestic waste. Many employment opportunities are generated in the biomass field, ranging from cultivation, collection, transportation, and processing, promoting rural development.

Demerits of biomass: Even though biomass seems an attractive source from a green and sustainable energy perspective, there are a few challenges that need to be addressed. For example, if we try large-scale cultivation of energy crops, then it may reduce the cultivation of food crops, limit land availability, cause deforestation, and water scarcity. It is noticeably clear that biomass is cleaner than fossil fuels, and its carbon dioxide generation is adjusted with the growth of next-generation plants, but large-scale and improper combustion can produce harmful pollutants such as particulate matter, nitrogen oxides, and volatile organic compounds. The efficiency of biomass systems is often much lower in comparison to solar or wind. Also, biomass plants have higher operation and maintenance costs.

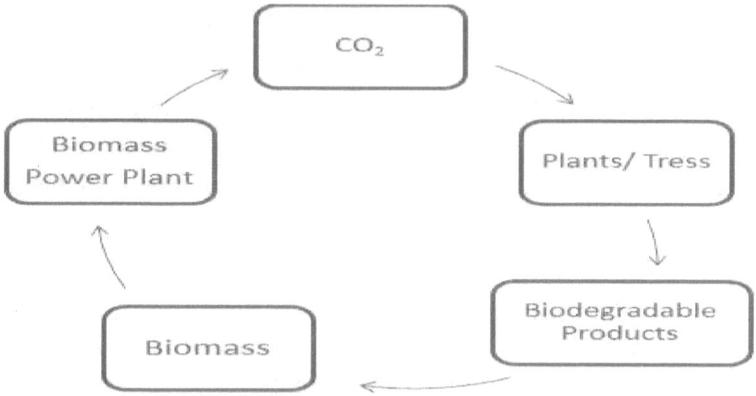

Figure 6.4 Life Cycle of a Typical Biomass

Role in Sustainable Development: The biomass sector directly supports many UN Sustainable Development Goals (UN SDGs).

SDG	Detail	Relevance
SDG 7	Affordable and Clean Energy	The biomass sector provides a decentralized renewable energy source.
SDG 8	Decent Work and Economic Growth	To help and promote rural employment.
SDG 12	Responsible Production and Consumption	Biomass directly helps in waste-to-energy production.
SDG 13	Climate Action	Biomass helps in the reduction of greenhouse gas emissions, promotes resilience, and local energy independence, thus contributing to carbon mitigation.

Table 6.4 SDGs and their Relevance Related to Biomass Energy

Technological Innovations: Modern gasification technologies can help in a more efficient way, which can be upgraded to biomethane using a combined heat and power system. Researchers are also exploring the production of lipids for diesel using microalgae, so that land for food crops and food security is not compromised for energy crops. Biochar, which is a carbon-rich by-product of the pyrolysis process (as mentioned earlier), is helpful in carbon sequestering and enhancing soil fertility.

> ***Activity***: *Visit the nearby village. List major sources of biomass. Can you suggest any novel idea to convert this waste into wealth? You may meet the collector or municipal commissioner about implementing the idea for the Swachh Bharat Abhiyan.*

Case Studies: With the incentives from the government, Sweden is the leader in the biomass energy sector, and derives more than 30% of its energy needs from biomass. They use biomass to get warmth in winter, using wood pallets and chips. On the other hand, Kenya promotes the installation of small-scale biogas digester tanks to produce biogas to be used for cooking, to reduce dependence on coal. India utilises sugarcane residues called bagasse to generate steam and electricity to ensure rural energy security.

6.3 Tidal and Wave Power

We are all surrounded by water in the oceans of the world. These vast oceans are constantly churning the water in the form of tides and waves. This energy can be harnessed to do useful work and generate electricity. As we are moving towards a paradigm shift from fossil fuels to renewable and non-polluting energy sources, marine sources offer great opportunities.

> ***Thinking Corner***: *Which fuel are you using in your home to cook food? Do you know, worldwide, 2.5 billion people still use biomass for cooking? Is it clean energy or a silent indoor health crisis? Think about gender impacts in rural biomass use. Do you help your mother cook food? How can you ensure better health for your mother through better fuel?*

Tidal energy refers to the capturing of energy from the rises and falls of ocean water due to the gravitational forces exerted by the sun and moon. These rises and falls are known as tides, which can be captured using either tidal barrages or tidal stream generators. The basic principle involved in tidal energy generation is either to capture potential energy from height differences between high and low tides or to capture kinetic energy from moving water using underwater turbines placed in fast-flowing tidal areas.

Wave energy refers to the energy generated using the waves that are created on the surface of the ocean water due to wind. Wave energy can be harnessed using point absorbers, oscillating water columns, and overtopping devices. Point absorbers are the floating structures that are equipped with hydraulic pumps and generators and move with the wave motion. In the case of oscillating water columns, air is compressed in a chamber using the wave to drive turbines, while in the case of overtopping devices, turbines are rotated through captured water in a reservoir.

Advantages and Disadvantages of Tidal Energy:

Advantages	**Disadvantages**
Tidal energy is highly predictable and reliable for energy generation.	Significantly high initial cost occurs due to the construction requirements, such as barrages.
Since water is denser than air, more energy can be generated using tidal energy than using wind energy.	Tidal power generation may impact the local environment of the marine ecosystem.
The lifespan of tidal power plants often exceeds 100 years; hence, they are cheaper in the long term.	Tidal energy plants can not be installed randomly. They are site-specific, where we get maximum tides depending on geographical conditions.

Table. 6.5 Advantages and Disadvantages of Tidal Energy

Advantages and Disadvantages of Wave Energy:

Advantages	Disadvantages
Since the ocean covers more than 70% of the Earth's surface, wave energy has vast energy generation opportunities.	Wave energy technologies have not yet fully flourished; they are still in the development and research stage.
Since most of the equipment for wave energy generation is submerged in water, they do not impact the aesthetic view.	Since these are submerged in water, their maintenance is a challenging task.
Wave energy may be used as a complement to other renewable energy sources like wind and solar.	They can impact the local marine ecosystem.

Table 6.6 Advantages and Disadvantages of Wave Energy

Comparison of Tidal and Wave Energy:

Tidal Energy	**Wave Energy**
Harnessed from the rise and fall of sea levels	Harnessed from the waves moving across the surface of the oceans.
The gravitational pull of the Moon and the Earth causes it.	The wind causes it.
Its intensity is affected by the movement of the Earth, the moon, and the Sun.	Its intensity is affected by the wind speed and direction, the intensity of sunrays, and the location.
Tidal energy is harnessed from the potential and kinetic energy.	Wave energy is harnessed from kinetic energy.
Tidal energy is a discontinuous source in which energy is generated for about 6 to 10 hours.	Wave energy is a continuous source of energy.
Tidal power plants require high initial costs but minimal maintenance costs.	Remarkably high initial cost, and still under research.

Table. 6.7 Comparison between Tidal Energy and Wave Energy

Global Deployment of Tidal and Wave Energy

Since oceans are vast, the global potential of tidal and wave energy is immense, but from the following table, it is clear that this immense treasure has not been properly utilised. It is estimated that globally tidal energy can provide 1,200 TWh and wave energy can provide 29,500 TWh per annum.

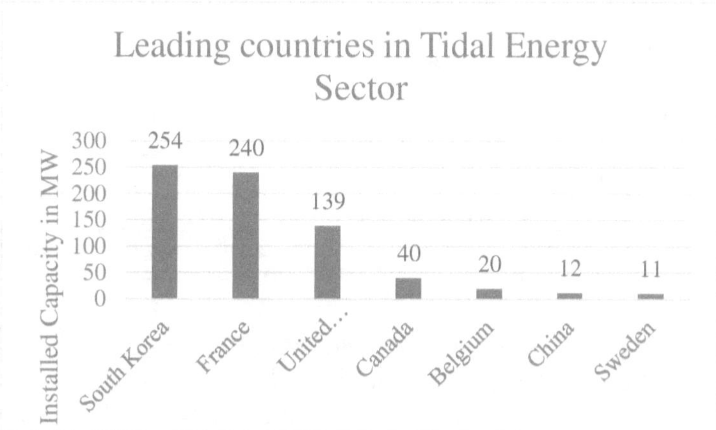

Figure 6.5 Worldwide Tidal Energy Generation

Sr. No.	Country	Tidal Power Plant Details	Installed Capacity	Remark
1	South Korea	Sihwa Lake Tidal Power Station	254MW	Operational
2	South Korea	Incheon Tidal Power Plant	1.32MW	Operational
3	South Korea	Undolmok Tidal Power Plant	1.5MW	Planned
4	South Korea	Saemangeum Reclamation Project	400MW	Planned
5	France	Rance Tidal Power Station	240MW	Operational
6	Canada	Annapolis Royal Generating Station	20MW	Operational

7	Canada	Fundy Ocean Research Centre	-	Test site
8	United Kingdom	MeyGen Project	6 MW	6 MW Operational 398MW Planned
9	United Kingdom	Bluemull Sound	0.5MW	Operational
10	China	Jiangxia Tidal Power Station	3.2MW	Operational
11	China	Baishakou Tidal Power Station	0.96MW	Operational
12	Russia	Kislaya Guba Tidal Power Station	1.7MW	Operational

Table 6.8 Worldwide Tidal Energy Plants (Operational and Planned)
**Global tidal energy potential estimates: ~1,200 TWh/year (source: IEA-OES, 2023).*

At present, one of the largest tidal stream projects at MeyGen in the United Kingdom has been installed. The LaRance tidal power project in France and the Agucadoura Wave Farm in Portugal are a few noteworthy commercial power plants.

Environmental Impact

It is clear that tidal and wave energy have immense potential and are yet to be explored. These plants have many advantages compared to conventional fossil sources. Despite lots of advantages, a few challenges ought to be considered. Since both tidal and wave energy equipment are in water, they may disrupt the marine ecology and alter the migration patterns of marine species. The salty water of oceans may cause early rusting; thus, proper care and maintenance are required. Coastal ecosystems may be impacted due to changes in sediment deposition by altered water flow.

6.4 Nuclear Fission and Nuclear Fusion

In the global warming context, it has become inevitable to search for, explore, and implement renewable energy sources and clean energy. In this context, when we move our attention towards nuclear power, it is often referred to as a contentious or dilemmatic topic due to its ability to deliver a huge amount of energy but also to become a source of mass destruction and catastrophe if misused.

.

Let us explore nuclear fission and nuclear fusion in detail in the following section. Nuclear fission is the process of breaking a nucleus of a heavy atom into two or more smaller atoms, along with the release of immense energy and a few neutrons. If these neutrons are allowed to strike heavy atoms, they will start a chain reaction. If uncontrolled, then it may liberate a huge amount of energy within a fraction of a second, turning into a nuclear bomb. However, if this chain reaction is allowed under controlled parameters, then we may generate electricity. Typically, heavy atoms like Uranium-235 or Plutonium-239 are used as a fuel. Generally, radioactive waste is produced during a nuclear fission reaction.

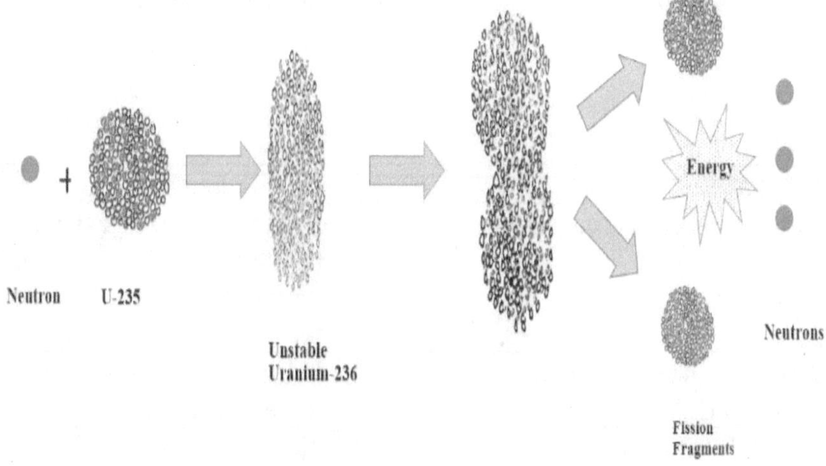

Figure 6.6 Liquid Drop Model to Demonstrate the Nuclear Fission

While nuclear fusion is the reverse process of nuclear fission. In the case of nuclear fission, two lighter nuclei are combined under certain conditions to form a heavy nucleus along with the release of immense energy. This reaction occurs naturally in all stars, including our sun. Nuclear fusion requires extreme pressure and temperature for its commencement. Little radioactive waste is generated in nuclear fusion.

Nuclear Fission:

Advantages	Disadvantages
Reliable base load power	The biggest problem is radioactive waste disposal.
High energy density 1 kg of Uranium≈24,000 MWh	High initial costs
Low greenhouse gas emissions	There is always a risk of accidents. e.g., Chernobyl, Three Mile Island, and Fukushima

Table 6.9 Advantages and Disadvantages of the Nuclear Fission Process

Nuclear Fusion:

Advantages	Disadvantages
Safe as no chain reaction occurs	Requires extreme pressure and temperatures to initiate the reaction
No long-lived radioactive waste is generated.	High research and development costs
Availability of abundant fuel Deuterium from seawater and Lithium for tritium breeding	No commercial fusion power plant has been established yet. Worldwide research is ongoing in different nations.

Table 6.10 Advantages and Disadvantages of the Nuclear Fusion Process

Comparison of nuclear fission and fusion processes:

Criteria	Nuclear Fission	Nuclear Fusion
Definition	A heavy nucleus is split into lighter nuclei	Lighter nuclei are combined to form heavy nuclei
Fuel	Uranium-235, Plutonium-239	Isotopes of hydrogen, like Tritium and Deuterium
Energy Generation	Lesser than nuclear fusion	Extremely high
Safety risks	Nuclear accidents, meltdown, radiation leakage	Less risky, but containment is yet challenging
By-products	Long-lived radioactive wastes	Minimum radioactive waste, Helium (non-toxic)
Use	To generate electricity in nuclear power plants.	In the experimental stage
Technology	Well established	Under development
Resource availability	Limited	Abundant
Temperature requirement	A few hundred degrees Celsius	Millions of degrees Celsius
Environmental Impact	Safe disposal of radioactive waste is always a challenging task.	Minimum waste, no radioactive wastes
Reaction Control	Fission reaction can be easily controlled with the help of neutron absorbers and moderators.	Extreme temperatures make control difficult.

Table 6.11 Comparison of Nuclear Fission and Nuclear Fusion Reactions

Nuclear Power and Sustainability:

Despite certain disadvantages of nuclear power, it is one of the cheapest and cleanest sources of energy. According to the IPCC, GHG emissions from nuclear fission power plants per unit are lower than those from solar PV and comparable to those from wind-based electricity generation. The following bar chart illustrates it well.

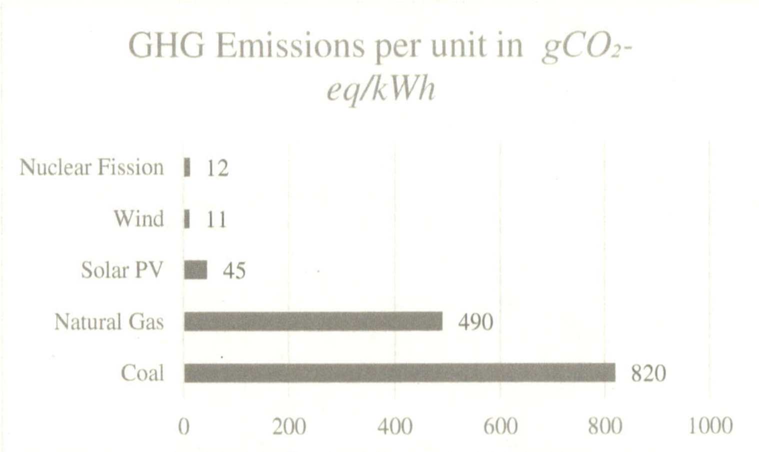

Figure 6.7 Greenhouse Gas Emissions of Nuclear Fission vs other fuels [3],[7]

Global Power Generation using Nuclear Fission:

Understanding the potential of nuclear power and keeping in mind the ill effects of climate change as a result of fossil fuels, most leading countries of the world have focused on the deployment of nuclear power. According to the International Atomic Energy Agency, as of 2024, more than 400 nuclear reactors are operational across the globe, contributing more than 10% of electricity. The USA is the leading country in this sector, having the largest number of reactors, while France's nuclear power supplies more than 70 % of its total electricity [3]. On the other hand, China and India have taken rapid steps to expand their capabilities in nuclear power. Following (*Figure 6.8*) is the list of leading countries along with their electricity generation in TWh for the year 2023.

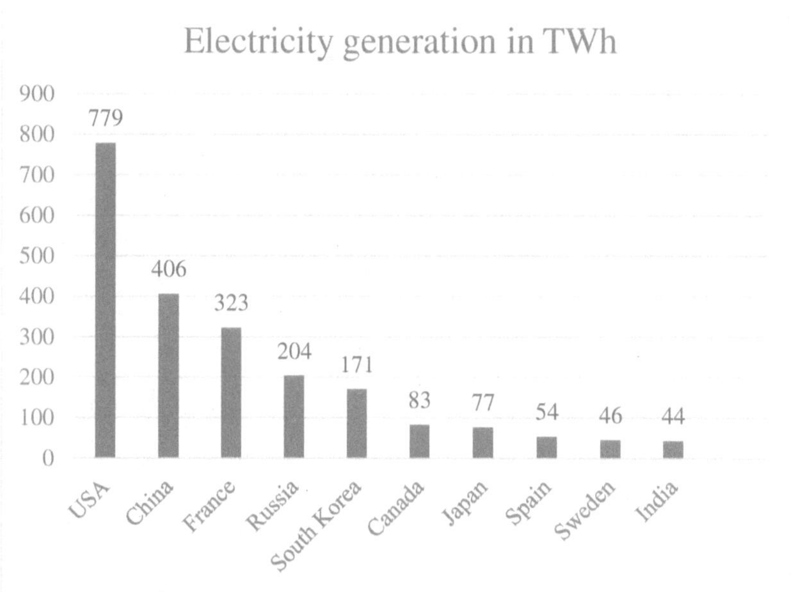

Figure 6.8 Global Power Generation using Nuclear Fission

Nuclear Fusion

As of 2025, no commercial nuclear power plant using fusion technology has been in operation despite its tremendous energy potential and other benefits. Nuclear fusion is under research at various stages. A collaborative research effort is planned at the world level under the umbrella of various organisations like ITER and SPARC. The International Thermonuclear Experimental Reactor (ITER) is the world's largest nuclear fusion research project based in France. It is a collaboration of 35 countries to achieve stable fusion energy.

K-STAR is another notable project in this area. In 2021, JET succeeded in achieving 59 MJ of sustained energy output; in 2022, NIF (USA) could demonstrate the net energy gain. ITER has an ambitious aim of full deuterium-tritium experiments by 2035 [38].

Nuclear Fusion reactors include passive safety systems, containment structures, and a cooling system. As said earlier, nuclear fusion reactors are inherently safe due to the lack of a chain reaction, no meltdown possibility, and limited fuel and meltdown risk. Nuclear waste from fission reactions has a life of 1000+ years, while that of fusion reactions has ~100 years. So, waste disposal in fusion reactions is easier in the case of nuclear fission.

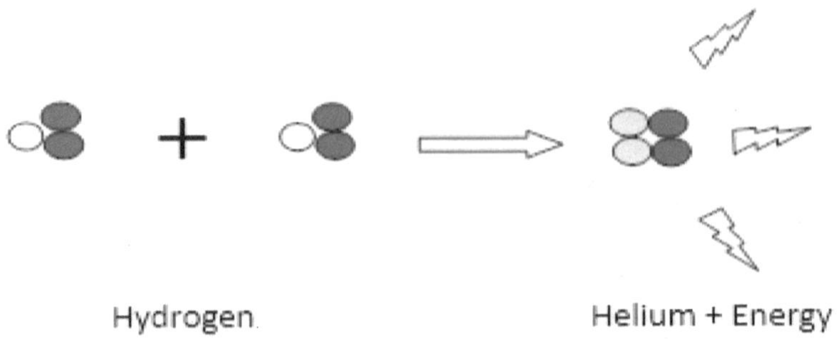

Figure 6.9 Nuclear Fusion

Role in Sustainable Development: The nuclear power sector supports many UN Sustainable Development Goals (UN SDGs).

SDG	Detail	Relevance
SDG 7	Affordable and Clean Energy	Nuclear power provides clean energy.
SDG 8	Decent Work and Economic Growth	Job creation in Engineering and Research.
SDG 9	Industry, Innovation, and Infrastructure	Nuclear energy directly promotes sustainable energy infrastructure.
SDG 13	Climate Action	Nuclear energy helps in the reduction of greenhouse gas emissions, promotes resilience, and local energy independence.

Table 6.12 SDGs and Nuclear Energy

To summarise, we may say that nuclear power plants are cheap in the long term but have a high initial cost, and there is always a risk of nuclear accidents. Comprehensive research is required for nuclear fusion. However, in nuclear fission reactions, Small Modular Reactors (SMRs) are playing a significant role in rural remote electrification and desalination plants in arid desert areas. Public perception is also a major factor to be considered while selecting the site, given the past nuclear accidents.

> ***Thinking Corner**: Experts have classified nuclear energy as green energy. Nuclear power emits no carbon, but it creates hazardous radioactive waste that lasts for thousands of years. Is it fair to solve today's problem by creating problems for our future generation? What's your opinion if you are going to take birth on this planet after 2000 years?*

6.5 Hydrogen Energy

Hydrogen is the most abundant element in the universe. Recent research on hydrogen has established it as the cleanest source of energy. Unlike fossil fuels, it does not release greenhouse gas emissions, but water! Hydrogen has emerged as a potential to establish itself as an ideal fuel for the future for various applications ranging from transportation, industries, and space. Hydrogen is a colourless, odourless, and non-toxic gas which is highly combustible. Hydrogen as an atom is not a source of energy, but it acts as an energy carrier. Its gravimetric density is 120 MJ/kg, which is three times that of petrol. Hydrogen can be transported after liquefaction. Hydrogen can also be used in internal combustion engines.

Hydrogen Production Methods: Hydrogen is abundantly available in nature, but it has to be extracted from hydrogen-rich compounds. Hydrogen can be classified based on its production methods as explained below:

- **Grey Hydrogen:** Grey hydrogen is produced through Steam Methane Reforming (SMR). This is the most widely used method today to produce hydrogen from natural gas (methane). However, this method is not environmentally friendly, as it emits a significant amount of CO_2.

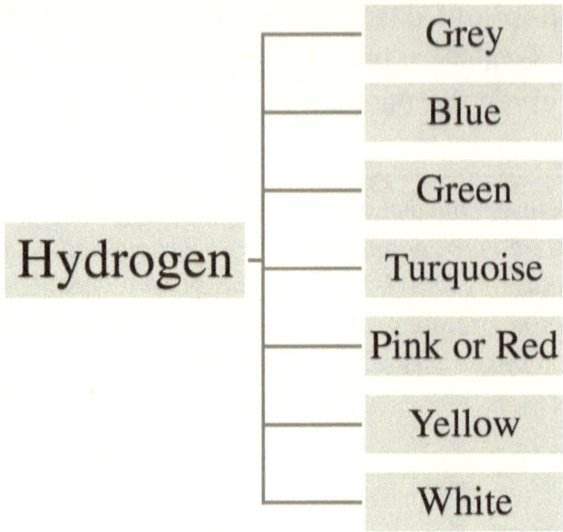

Figure 6.10 Types of Hydrogen

- **Blue Hydrogen:** This method is an advanced form of the above method. This method also uses SMR but with the addition of Carbon Capture and Storage (CCS) to absorb emissions. Since an additional facility for carbon sequestration is required, this method is expensive. Dependence on fossil fuels is one more disadvantage of this method.
- **Green Hydrogen:** This is the ideal method of hydrogen production, but it is very costly. Due to prohibitive costs and limited infrastructure, less than 0.1% of total hydrogen is produced using this method *(source: IEA 2024 report)*. Green hydrogen is produced using electrolysis, in which water is split into oxygen and hydrogen. Electricity required for splitting water is derived from renewable energy sources like solar and wind. Hence, no greenhouse gas emission is associated with this process.
- **Turquoise Hydrogen:** It is produced from methane pyrolysis, which gives hydrogen and solid carbon. This method is considered to be better because solid carbon is easier to store than the gas CO_2. This method has not yet been widely explored.
- **Pink or Red Hydrogen:** Its source is electrolysis powered by nuclear energy. This method can provide reliable, stable

hydrogen and has a low carbon footprint; however, there will always be a risk of nuclear waste and safety concerns.

- **Yellow Hydrogen:** In this case, hydrogen is produced using electrolysis, but electricity is used from the grid, which may be a mix of fossil and renewable energy sources. The carbon footprint in this case depends on the grid mix at that particular time.
- **White Hydrogen:** This hydrogen is also known as Geologic Hydrogen, as it is naturally occurring underground hydrogen. This hydrogen has not yet been commercially exploited. This may become a future clean source of hydrogen if found in enough amounts.

Applications of Hydrogen

Hydrogen is universally available, making it a versatile source for a wide range of applications. Research is being conducted to explore the potential of hydrogen for various applications. At present, hydrogen is widely used in the transportation sector. Fuel cell electric vehicles are powered by hydrogen, have a longer range, and can be refuelled faster compared to battery-operated vehicles. This makes them a better choice for long-range heavy-duty trucks, buses, trains, and even planes. In industries, hydrogen can replace coal in the manufacturing of steel and ammonia production. Electricity can also be generated using hydrogen, particularly during the peak demand hours. It can also serve as a storage medium in place of batteries. We may convert surplus renewable energy from solar and wind into hydrogen and use it in peak demand.

Benefits of Hydrogen Energy

Hydrogen can prove to be a boon if we can find a cheaper way to produce hydrogen on a mass scale. Hydrogen fuel can help us achieve the target of net-zero emissions. Green hydrogen can eliminate carbon dioxide as it uses only renewable energy sources. If we replace diesel and petrol-based engines with hydrogen engines, then air quality will also be improved, leading to better health for all animals, including humans, due to the reduction of NO_x, SO_x, and particulate matter. Introduction of hydrogen for energy backup and energy production can also increase the diversification of energy sources and the reliability of the grid.

Millions of jobs may be created in manufacturing, construction, and the service sector.

Technological Challenges: Although hydrogen energy looks so attractive, there are multiple challenges yet to be addressed. Green hydrogen will be the ultimate choice due to its minimal carbon footprint, but the production cost of hydrogen using electrolysis is very high. We have already mentioned that hydrogen can easily be stored for a long time and can also be transported to distant places, but the low density of hydrogen makes its storage a challenging task. Hydrogen must be compressed to 350 to 750 bars for liquefication at -253 °C, which is energy-intensive and costly. Significant infrastructure has to be developed for electrolysis, fuelling stations, and pipelines. Even when hydrogen is a non-toxic gas, it is a highly combustible gas with a wide ignition range. So, leak detection and its containment require special protection, public awareness, training, and safety concerns.

Global Initiatives: Integration of hydrogen to generate green electricity or to reduce the carbon footprint is essential, given climate change and global warming. Hence, almost all leading nations have started to invest in the fuel of the future, that is, hydrogen. The European Union aims to install 40 GW of domestic electrolysis capacity by 2030 [10]. Japan plans to become the world's first hydrogen economy! India has launched the National Green Hydrogen Mission to achieve 5 MMT annual production of hydrogen by 2030 [12]. India's transport minister, Mr Nitin Gadkariji, has been strongly advocating for the hydrogen transport system in India. Australia is planning to produce and export hydrogen.

> ***Thinking Corner**: Dream of a future! Suppose you have found a very easy novel way to convert water into hydrogen. Now, every household, business, and industry may have its own energy generator using hydrogen storage (just like today's diesel generator). There won't be any need for large dams, nuclear power plants, no more transmission lines……*
>
> *What else may be the change in our lives?*

Role in Sustainable Development: The hydrogen energy sector supports many UN Sustainable Development Goals (UN SDGs).

SDG	Detail	Relevance
SDG 7	Affordable and Clean Energy	The hydrogen energy sector provides clean energy.
SDG 8	Decent Work and Economic Growth	Job creation in Engineering and Research.
SDG 9	Industry, Innovation, and Infrastructure	Hydrogen energy directly promotes sustainable energy infrastructure.
SDG 11	Sustainable Cities and Communities	The hydrogen energy sector helps in creating sustainable cities and communities by providing cleaner air and mobility.
SDG 13	Climate Action	Hydrogen energy helps in the reduction of greenhouse gas emissions, promotes resilience, and local energy independence.

Table 6.13: SDGs and Hydrogen Energy

-:EXERCISE:-

Section A: Short Answer Questions

1. List the main components of geothermal power plants.
2. Mention the advantages of geothermal energy in cold countries.
3. From an Indian perspective, list possible sources of biomass.
4. State the environmental issue associated with tidal power plants.
5. Compare nuclear fusion and nuclear fission reactions.
6. List major nuclear accidents of nuclear power plants.
7. List types of hydrogen.
8. Compare nuclear fusion and fission reactions.
9. List technical challenges concerning hydrogen storage and transport.
10. State the advantages of biogas plants.

Section: Long Answer Questions

1. Discuss the advantages and challenges of geothermal energy as a renewable resource.
2. Briefly analyse the environmental impact of geothermal power plants.
3. Compare tidal energy and wave energy.
4. State how geothermal energy contributes to fulfilling the UN sustainability goals.
5. Describe how biomass can help improve life in rural India.
6. Evaluate the scope of tidal and wave power at the global level.
7. Compare various processes of hydrogen production. Do you feel that hydrogen is a fuel for the future?
8. Analyse the possible role of nuclear energy (fusion and fission) in achieving the UN SDGs.
9. Explain the risks associated with nuclear fission.
10. Explain the nuclear fusion process and challenges preventing its commercial use.

Section C: Multiple Choice Q and A

1. The primary source of geothermal energy is:
   a) Wind currents
   b) Radioactive decay inside the Earth
   c) Solar radiation
   d) Ocean tides

2. ....type of geothermal power plant uses direct steam from a hot reservoir.
   a) Wet Steam　　b) Dry Steam
   c) Dry moisture, Steam　　d) Superheated Steam
3. The geothermal power plants require a temperature range for electricity generation.
   a) Below 50°C　　b) 50°C - 150°C
   c) Above 150°C　　d) Any temperature
4. .... country uses geothermal energy to warm over 90% of its homes.
   a) Iceland　　b) Kenya.
   c) India　　d) Sweden

5. .... Geothermal power plant type is suited for lower temperature sites (100 - 170°C)?
   a) Flash Steam Plant b) Dry Steam Power Plant
   c) Binary Cycle Plant d) None of the above
6. The following is NOT an advantage of geothermal energy.
   a) Minimal land footprint b) Low water usage
   c) Low temperature d) High GHG
7. Enhanced Geothermal System uses the principle of:
   a) Injecting water into fractured rocks to extract heat
   b) Using solar heat to boost geothermal energy
   c) Using wind to increase the geothermal temperature
   d) Combining geothermal and fossil fuels
8. Biomass energy has almost zero carbon footprint because:
   a) It produces no CO_2 emissions
   b) CO_2 produced is offset by plant growth
   c) It produces methane
   d) It uses fossil fuels
9. A biomass conversion method to produce biodiesel is:
   a) Thermochemical Conversion
   b) Biochemical Conversion
   c) Direct Combustion
   d) Transesterification
10. A disadvantage of large-scale biomass cultivation is:
    a) Increased food crop production
    b) Land availability issues
    c) No effect on water resources
    d) Zero emissions
11. .... country is a global leader in biomass energy and derives over 30% of its energy needs from biomass?
    a) India b) Kenya
    c) Sweden d) France
12. Tidal energy is caused by:
    a) Ocean currents
    b) Gravitational forces of the Moon and Sun
    c) Wind patterns
    d) Earthquakes
13. A major disadvantage of tidal energy plants is:
    a) Unpredictable energy output

b) High environmental impact on the marine ecosystem
c) Low lifespan
d) Low initial capital cost

14. Wave energy is harnessed mainly from:
a) Tidal differences in sea levels
b) Surface ocean waves caused by wind
c) Ocean currents caused by temperature differences
d) Geothermal heat under the ocean floor

15. ...... country has the largest operational tidal stream projects.
a) India b) South Korea
c) United Kingdom d) Morocco

16. The energy content of 1 kg of Uranium-235 in nuclear fission is:
a) 2,400 MWh b) 24,000 MWh
c) 2,40,000 MWh d) 240 MWh

17. An accident which is NOT associated with nuclear fission reactors is:
a) Fukushima b) Chernobyl
c) Three Mile Island d) Deepwater Horizon

18. Which of the following gases is a by-product of nuclear fusion?
a) Carbon dioxide b) Helium
c) Methane d) Nitrogen

19. ITER aims to be fully operational by .... year.
a) 2025 b) 2030 c) 2035 d) 2040

20. A nuclear fusion project that successfully achieved 59 MJ of sustained energy output in 2021 is:
a) SPARC b) JET c) NIF d) K-STAR

21. Which is a true statement?
a) Fusion waste lasts longer than fission waste
b) Fusion waste lasts about 100 years; fission waste lasts over 1000 years
c) Fusion waste lasts over 1000 years; fission waste about 100 years
d) Both last equally long

22. .... is NOT a source of hydrogen production.
a) Electrolysis powered by nuclear energy
b) Steam Methane Reforming (SMR)

c) Methane pyrolysis
d) None of the above

23. Turquoise hydrogen is environmentally better than grey hydrogen because:
a) It produces no CO_2 gas but solid carbon instead
b) It is the cheapest production method
c) It uses renewable energy only
d) It produces more hydrogen per kg of methane

24. Hydrogen fuel cell vehicles have .... advantages over battery electric vehicles.
a) Lower refuelling time and longer range
b) Use fossil fuels as backup
c) No need for hydrogen infrastructure
d) Lower initial cost

25. .....Sustainable Development Goal is linked to job creation in the nuclear energy sector.
a) SDG 7 –Climate Action
b) SDG 8 – Decent Work and Economic Growth
c) SDG 9 – Industry, Innovation, and Infrastructure
d) SDG 13 – Affordable and Clean Energy

26. A temperature requirement for initiating nuclear fusion is:
a) A few hundred degrees Celsius b) Room temperature
c) Millions of degrees Celsius d) Around 1000 °C

27. CCS stands for .... about blue hydrogen.
a) Carbon Capture and Storage
b) Chemical Carbon Separation
c) Controlled Combustion System
d) Carbon Cleaning Staff

28. .... country currently has the largest number of operational nuclear reactors.
a) France b) China c) USA d) India

29. France generates............... electricity from nuclear power.
a) 30% b) 50% c) Over 70%
d) About 10%

30. The......... method of hydrogen production has high CO_2 emissions.
a) Grey hydrogen b) Pink hydrogen
c) Green hydrogen d) Blue hydrogen

Answer Keys:

1. B, 2. B, 3. C, 4. D, 5. C, 6. A, 7. C, 8. B, 9. D, 10. B, 11. C, 12. B, 13. B, 14. B, 15. C, 16. B, 17. D, 18. B, 19. C, 20. B, 21. B, 22. D, 23. A, 24. A, 25. B, 26. C, 27. A, 28. C, 29. C, 30. A

भुञ्जीत यावदर्थं तु न भुञ्जीतातिलम्पटः।
भोगेष्वनभिसङ्गेन तथा जीवेत पण्डितः॥

Mahābhārata, Śānti Parva 262.6

One should consume only as much as is necessary, not indulge in excess.

Today, we say *"reduce energy consumption,"* while our ancestors said, *"avoid excess, consume only as needed, live in harmony with nature."*

Chapter 7

GREEN INFORMATION and COMMUNICATION TECHNOLOGIES (ICT)

Learning Outcomes:

After reading this chapter, the learner will be able to:

1. Understand the concept of Green ICT in terms of sustainability.
2. Assess the impact of ICT infrastructure on the environment.
3. Discuss the ways to reduce the carbon footprint of ICT infrastructure.
4. Analyse the life cycle management of ICT gadgets to reduce e-waste.
5. Explore the role of ICT in promoting sustainability.

7.1 Introduction

Can you imagine a world without gadgets? Try to live a day without mobiles and the internet! Seems exceedingly difficult, correct? These devices have become an integral part of our lives. We are surrounded by gadgets from morning till late at night. Our mind is constantly bombarded with lots of information, and our body is completely drenched in an ocean of microwaves for the entire day and night. Lots of energy is required to install, operate, maintain, and conduct research in this sector. Information and communication technologies are popularly known as ICT. It includes digital devices like mobile phones, tablets, laptops, computers, and networks. The digital revolution has transformed our lives, making it possible to be in touch with people, to socialize, to express our thoughts, and learning. However, ICT usage has a significant environmental impact. Giant Giga Data Centers, network infrastructure, and millions of electronics, IoT, and computational devices/gadgets use a huge part of electricity and create tons of e-waste annually. On the other hand, ICT can also help sustainability in many other sectors.

Green ICT refers to the use of these devices in such a way as to reduce the environmental impact to a minimum. In this chapter, we shall try to focus the discussion on understanding the meaning of Green ICT, its environmental impact, and strategies to reduce the carbon footprint of ICT. We shall also focus on the recycling of ICT devices at the end of their useful life and the role of ICT in sustainability.

7.2 Green ICT

Green ICT refers to the concept of designing, building, and using ICT resources in such a way that they consume less energy, release minimal greenhouse gas emissions, and create minimal waste during and at the end of their useful life. ICT includes physical devices like phones, smart TVs, computers, and the networks that connect them, as well as software and services that run on these devices. The primary goal of Green ICT is to reduce the ecological footprint of the digital world. Green ICT also involves the effectiveness and efficiency of physical devices, as well as software and services. For this, standards and metrics have been framed to measure sustainability. For example, Power Usage Effectiveness (PUE) is a standard to compare the total energy used by a facility and the energy used by computing equipment. On the other hand, carbon footprint calculators are used to calculate the total greenhouse gas emissions by ICT activities. Engineers measure the energy use and emissions to improve the design.

Worldwide, two key metrics are recognized to evaluate the data centers' efficiency. They are Power Usage Effectiveness (PUE) and Data Center Infrastructure Efficiency (DCiE). These metrics quantify how "green" a data center is. PUE refers to the ratio of the total power requirement of a data center to the power required by IT (computing, storage, networking) equipment only. A typical value of PUE is 2, which shows that other infrastructure consumes power as much as IT equipment. Leading companies like Google, Microsoft, and Amazon have been able to achieve PUE values for their centers as low as 1.32. The following table sheds light on various energy efficiency metrics.

Metric	Ratio of	Typical Value	Good Target	Description
Power Usage Effectiveness (PUE)	Total Data Centers' power/ IT equipment. power	2	Less than 1.3	It evaluates the overall efficiency. Lower value is better.
Data Center Infrastructure (DCiE)	IT equipment. Power/Total Data Centers	0.5	More than 0.625	It reflects the proportion of power consumed by

	power. It is the inverse of PUE.			IT equipment.
Data Center Efficiency (DCE)	HVAC power/IT equipment. power	0.5	More than 0.625	It often relates to cooling efficiency compared to IT load. A lower ratio is better.
Data Center Energy Productivity (DCeP)	NA	NA	NA	It is a proposed metric to quantify energy efficiency.
Green Grid Productivity Indicator	NA	NA	NA	It is a proposed metric to quantify energy efficiency.

Table 7.1 Various metrics to measure energy efficiency of data centers.

A typical DCE value for a data centre is 0.5, but the target should be above 0.625. Developing new metrics is essential, other than DCE and PUE, to better understand the carbon footprints of ICT equipment. Any data centre's energy use will comprise IT equipment, cooling infrastructure (HVAC), and other essential facilities like lighting, water pump, lift, etc.

To summarize, we may say that Green ICT is an approach to consider sustainability at every step to conserve energy and resources, and to minimize the damage to the ecology.

7.3 Environmental Impact of ICT

Year	Total ICT Energy Use in TWh	Data Centre Energy Use in TWh	Data Transmission	Remark
1975	30	N.A.	N.A.	Early computing era
1980	50	N.A.	N.A.	Mainframe expansion
1985	80	N.A.	N.A.	Emergence of personal computers
1990	120	10	5	Network infrastructure started
1995	200	25	10	Internet expansion started
2000	350	50	20	Dot-com boom
2005	500	80	40	Broadband expansion
2010	700	120	80	Rise of cloud computing
2015	800	200	120	Stable growth
2020	900	270	200	Dominance of mobile internet
2022	950	300	260	AI and video streaming rise

Table 7.2 Global ICT energy consumption.
(Source: IEA and IPCC reports)

Information and communication equipment and services consume a significant amount of energy worldwide. It is estimated that as of 2022, the ICT sector consumed approximately 4% of global electricity, which is equivalent to 1,000 TWh annually [3]. Future projections suggest that the ICT sector's energy demand may rise to

15% of global electricity by 2030 due to increased data traffic and AI penetration in daily life. India's IT sector is expanding very rapidly; at present, it consumes 0.5% of the country's electricity, but it may rise to 3% by 2030. The following table indicates the global ICT energy consumption.

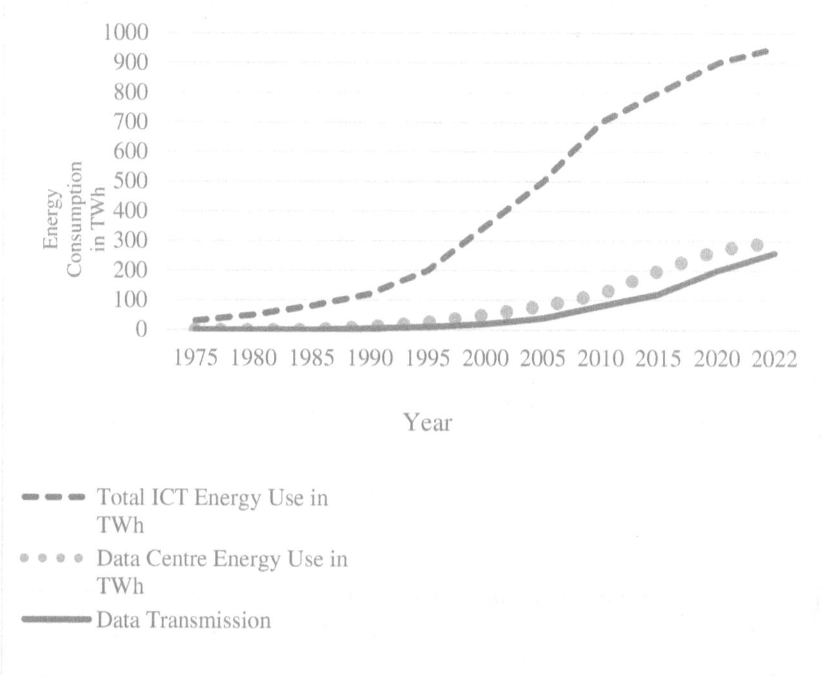

Figure 7.1 Visual representation to understand the consumption and demand pattern for energy in the ICT sector at the global level. (Source: IEA and IPCC reports)

> ***Thinking Corner**: The Paradox of Progress!*
>
> *Can today's ICT gadgets solve environmental problems without creating new ones?*
>
> *According to you, what's the balance?*

The following graph, Figure 7.2, helps us to visualize the pattern of energy demand in India for the ICT sector.

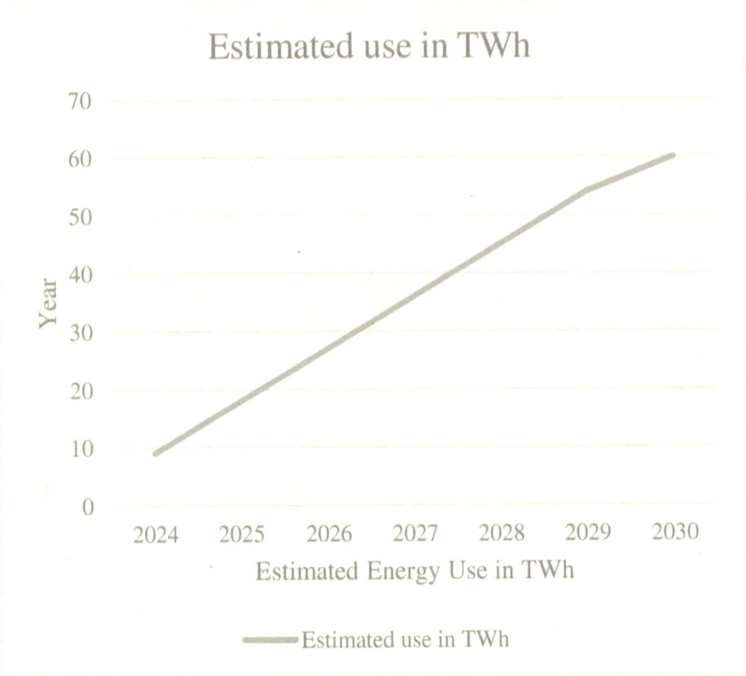

Figure 7.2 Visual representation to understand the consumption and demand pattern for energy in the ICT sector for India. (Source: NASSCOM and IEA)

From the above two representations, it is very clear that energy demand for the ICT sector is increasing rapidly due to the widespread use of AI, IoT, video streaming, data centres, and data transmissions. Because of high energy use and mass production of ICT equipment, they significantly contribute to environmental degradation. The two main areas of impact are greenhouse gas emissions and electronic waste. Let's explore these in detail.

7.3.1 Carbon Emissions from ICT

We have already discussed that the ICT sector is heavily consuming electricity. Since in most countries, electricity is generated from fossil fuels, it is indirectly responsible for greenhouse gas emissions.

The sources of these emissions include data centers, networks, user devices, and manufacturing materials.

- **Data Centers:** Data centers provide a facility to house servers for cloud computing and internet services. They run 24x7 and require power for ICT devices, lighting, and the cooling of servers. You will be surprised to read that a typical large data center may consume power in the range of 10 to 100+ MW! This may be around 87.6 million kWh/year to 876 million kWh/year. Typical floor space may be around 1,00,000 sq. ft to 5,00,000 sq. ft. All Modern data centers try to reduce their carbon footprint by using efficient techniques, better cooling design, and relying on renewable energy sources. A typical Google data center consumes 20-50 MW, with PUE 1.10. Facebook data centers in Prineville, Oregon, use around 30 MW, while Microsoft data center uses 200+ MW per facility.

- **Data Networks and Telecommunications:** A Data network connects various devices like cellular towers, optical fiber cables, switches, and routers. A single cellular tower may consume 1- 5 kW of electricity for radios, antennas, and processing. These towers are in lakhs across the globe; hence, their power consumption reaches a significant amount. Advanced technologies are more energy efficient and reduce the carbon footprint.

- **User Devices:** We use several gadgets and equipment such as smartphones, laptops, tablets, personal computers, and monitors. These devices use power in tens of watts. We may think that small devices like smartphones hardly consume 60 kWh per annum, so why worry about their electricity, correct? Do you know the total number of smartphones in the world? It's about 6.9 billion! When we calculate the total power consumption of all these smartphones, it reaches approximately 414 TWh per annum! It is comparable to 80% of Germany's or 100% of Saudi Arabia's electricity demand!

- **Manufacturing and Materials:** The production of ICT

devices involves various processes like mining, refining, plastic production, manufacturing, assembling, packaging, and marketing. It is believed that for consumer electronics like smartphones, manufacturing can contribute the majority of the device's life-cycle carbon footprint rather than the actual use. We can better understand this concept from the following table. We have assumed 100% recycling of equipment. It can be inferred from the following table that smaller devices have a higher carbon footprint during manufacturing, while in larger equipment, this part is significantly reduced due to a longer lifespan.

Device	Production (materials and assembly)	Electricity Usage	End-of-life (Assuming 100% recycling)
Smartphone	~70%	~30%	~0%
Laptop	~50%	~50%	~0%
Desktop PC	~40%	~60%	~0%
Server rack	~30%	~70%	~0%

Table 7.3 Approximate share of GHG for different ICT devices

The following table gives a rough comparison of power consumption and carbon emissions per annum.

Device	Production (materials and assembly)	Annual Electricity Usage in kWh	Annual CO_2 kg (Assuming 0.5 kg CO_2 per kWh)
Smartphone	~5 W (charging)	~10	~5 kg
Laptop	~50 W	~500	~250 kg
Desktop PC	~200 W	~3,000	~1500 kg
Data Center Rack	~10 kW	~87,600	~43,800 kg

Table 7.4 Approximate Annual CO_2 generation in kg *for different ICT devices*

A few Important Facts:

- Greenhouse gas emission depends on the source of electricity generation. For example, if electricity is generated from coal, then each kWh emits 1 kg CO_2, for natural gas it's 0.4 kg CO_2, while for renewable energy sources like wind, solar, and hydro, it's negligible. Thus, the carbon footprint of ICT is dependent upon the energy mix of the local grid.

- Advancements in processor technology, solid-state devices, and natural cooling have reduced the energy requirement of data centers. For example, Google has reported PUE around 1.1 in its few data centers, which is remarkably close to the ideal value of 1.0.

Overall, the carbon footprint of ICT devices is much smaller in comparison to sectors like industries and transportation, yet it is better to switch to renewable energy sources for electricity requirements.

7.3.2 E-Waste from ICT

E-waste or electronics waste refers to the discarded, obsolete, unused, thrown, broken electronic devices and components. It may include smartphones, televisions, printers, cables, network switches, servers, computers, laptops, and many more. E-waste poses a big environmental challenge as it has many hazardous substances and minerals like mercury, lead, and plastics. If this waste is not properly disposed of/recycled/managed, then it may contaminate soil and water. This contamination may pose serious health issues to us.

We are seeing the revolution of communication and the internet, but at the same time, the production of these gadgets is also rising at an exponential level. These ICT devices become e-waste after their end-of-life, which is their usable life. As per a rough estimate, we are generating 7 to 8 kg of e-waste per person globally.

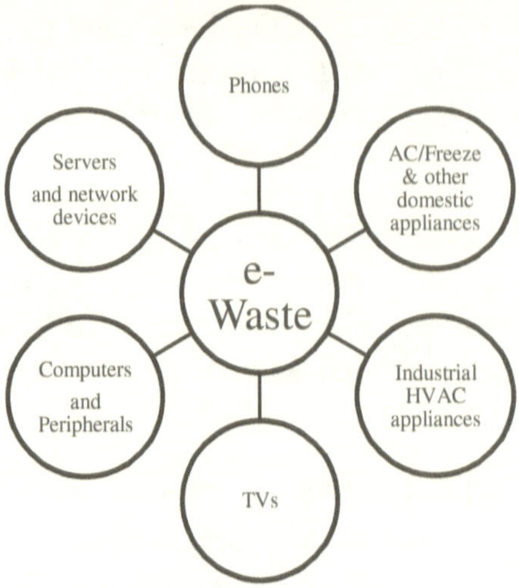

Figure 7.3 Types of e-wastes

According to the Global E-Waste Monitor (2022 Report), globally, ~59.4 million metric tons of e-waste were generated in 2022. It may reach around 82 million by 2030. It is due to the rapid rate of replacement of certain gadgets like smartphones and laptops. Unfortunately, only 17.4 % of e-waste is formally collected for recycling; the rest is either discarded in landfills or dumped in low-income countries [39]. E-waste should be properly recycled through dismantling, separating materials, and recovering metals through chemical processes. This is often costly, but it gives the benefit of reusing the same materials, conserving valuable natural resources, and reducing pollution.

It may be noted that large appliances like fridges, TVs, and

***Thinking Corner**: The Paradox of Progress!*

More than 62 million tonnes of e-waste are generated globally per annum. Millions of people are exposed to toxic heavy metals like lead, mercury, and cadmium, leading to the death of ~ 1 million people every year. Aren't we all responsible for this crisis through our careless consumption and irresponsible e-waste disposal? What may be the solution ?

washing machines account for most of the weight of e-waste, but a concentration of valuable metals is higher in small items like smartphones and laptops. Interestingly, if we recycle one ton of mobile phones, then we may recover more than 100 times more gold than one ton of gold ore! Green ICT emphasizes the reduction of e-waste by extending lifetimes, promoting repair, reuse, and recycling to minimize damage to the environment.

7.4 Green ICT Lifecycle

An overall carbon footprint of ICT equipment may be reduced by improving efficiency along with its design, manufacturing, extended responsibility, and end-of-life offers.

7.4.1 Sustainable Design**:** The design phase of any equipment is particularly important as it affects the whole life of the equipment. Design should be focused on creating products with fewer resources, energy efficiency, longer life, easier to recycle, and minimal harm to the environment.

a. Material selection: Choose those materials that have a lower environmental impact. Prefer materials that are abundant in nature or renewable. Each material is associated with GHG emissions for its processing, so the wise choice of material reduces the carbon footprint of ICT equipment. For example, the use of recycled plastics, recycled metals, bioplastics, and non-toxic materials can reduce the carbon footprint.

b. Modularity and Upgradability: ICT equipment should be designed in such a way that worn, burned, or damaged components can be easily replaced or upgraded. This will save the equipment from being discarded. For example, removable batteries and removable memories can extend the life of a device.

c. Energy Efficiency: Use low-power-consuming components to conserve energy. Implementation of sleep modes and optimized layout of circuit boards can also contribute to power saving during useful life. Energy-saving electronics should be used for those devices that run continuously, like routers and network switches.

d. Product Packaging: Waste reduction can also be achieved by designing minimal and recyclable packaging material. Packaging design also affects shipping emissions (smaller boxes mean more items per truck).

e. Lifecycle Assessment (LCA): A formal LCA during the design stage can find the component with maximum environmental impact. Replacement of such components can improve overall efficiency throughout the ICT life, or it can minimize the environmental impact post-useful life.

Strategy	Purpose	Example
Material	Lower Carbon footprint or toxicity	Use recycled plastics, avoid toxic flame-retardants.
Durability and Repair	Extend product life	Allow easy replacements, robust moving parts
Power Saving Features	Reduce Energy Consumption	Include hardware-based sleep modes. Use efficient electronics components.
Modular Structure	Ease of repair and upgradability	Screw in parts, standard modules

Table 7.5 E-waste reduction strategies for ICT devices

Many companies have started to implement these strategies, such as the use of recycled metal and materials, dropping PVC and brominated flame retardants, and adopting modular architecture. Sustainable design balances performance and cost while minimizing environmental damage.

7.4.2 Manufacturing Practices: A manufacturing process involves turning raw materials into products. The green manufacturing process involves:

- The use of clean energy sources like solar and wind.
- Waste reduction through lean manufacturing processes.
- Use of recycled water.
- Reducing the use of toxic and harmful chemicals and solvents. Ensuring the release of wastes only after proper treatment to reduce pollution.
- Production close to the consumption point to reduce transportation.
- Obey ethical and moral standards to accept social responsibility to conserve the environment.

7.4.3 Energy Efficiency: Once devices are manufactured, transported to the site, installed, and assessed, they are ready to use. This phase often consumes the largest amount of energy in its lifecycle. So, proper strategies should be incorporated to reduce the energy consumption in operation. This can be improved by using higher efficiency ICT equipment, for data centers, use low-PUE designs, replace old servers with the latest energy efficient servers, and use energy efficient software to cut lengthy and redundant codes. Enable built-in energy saving modes, use advanced internet protocols to optimize data transmission using data compression, use energy-efficient Ethernet like 802.3az, and adjust cooling at data centers.

7.4.4 End-of-Life Management

Every ICT device has a fixed lifetime to work properly. After that, it may be obsolete in the market due to the availability of more advanced devices (sometimes even at cheaper prices!), non-functioning, breakdown, or due to ageing. It is particularly important to dispose of these devices properly to treat harmful toxic minerals, to recollect valuable metals, and rare Earth minerals.

- Reuse and refurbishment: old devices should be donated to needy people. For example, a company may acquire new, upgraded computers, but should donate old working computers to nearby schools. These computers can play an excellent role for school students, as they don't need to run high-end software.

- Safe disposal: Batteries and CRT monitors have heavy metals and should not be dumped in landfills. Smartphones also have many valuable minerals and batteries. Safe disposal of ICT is very much necessary to save our water and soil from being polluted. Most mobile batteries have lithium, cobalt, nickel, manganese, electrolytes, and other heavy metals in trace amounts. They can leach into water and affect the aquatic ecosystem, accumulate in fish, and enter the human food chain. This can result in tissue damage, reproductive failure, and death of fish and other amphibians. Battery chemicals can also seep into the groundwater and affect human health. So, it's especially important to be aware of the safe disposal of ICT gadgets.
- Extended producer responsibility (EPR) requires manufacturers to take back the product after the end of its useful life. This mandates the producers to design products that are easier and cheaper to recycle.

The sustainability loop is closed by proper end-of-life. If used/broken/obsolete ICT devices are reused, refurbished, repaired, or recycled, then raw materials stay in circulation, and less mining will be needed for virgin resources.

7.5 ICT as an Enabler of Sustainability

Beyond making ICT itself green, we can use ICT to increase the sustainability of many other sectors. ICT4S (ICT for Sustainability) is a concept of using ICT to make sectors more sustainable. Modern digital and IoT devices help to reduce energy consumption in transportation and agriculture, automate the process, and promote water conservation in buildings and industries. Let us explore a few superficially.

- Smart buildings and cities can increase energy efficiency, water conservation, and water harvesting. It also includes waste management, real-time temperature optimization based on smart thermal sensors by monitoring occupancy.

Green buildings can become more efficient by integrating the Building Management System (BMS) to control the mechanical, electrical, and electronic devices from a common network to increase the overall efficiency of a building. It includes ventilation, air-conditioning, lighting, power system, fire safety, lift management, water pumps, and security.

- Smart grids use digital communication to automatically adjust or shift the load based on demand. For example, momentary dimming streetlights during heatwaves, and auto-cutting non-essential loads to meet the important peak load.
- Traffic and transport management can become smart by using adaptive traffic lights to reduce idling and congestion (thus saving fuel). Ride-sharing apps may use AI-enhanced maps to select the least congested route to reduce fuel consumption.
- Electric and autonomous vehicles are the future of transportation. EVs rely on ICT systems for battery management, smart charging networks, all of which can be controlled and optimized with the help of ICT.
- ICT directly reduces transportation due to its ability to arrange online video conferences and meetings. This saves time and fuel.
- The agriculture sector is also a beneficiary of ICT to decide the time of sowing, to receive weather forecast data about monsoon and heat waves. ICT-controlled drones can help to spray pesticides on crops.

7.6 Challenges

Despite ICT looking so attractive, a way to live in the future, its implementation faces many challenges. Upgrading to greener and more efficient technology always requires significant upfront costs. Many times, the payback period is long enough to make it economically unviable. So, ordinary people and small business owners hesitate to implement it. Individuals and organizations are not aware of the environmental damage due to irresponsible behavior regarding the e-waste of ICT gadgets. Low-income countries are lacking the infrastructure, internet, data centers, network, cloud services, and smart solutions to reap the fruits of ICT. Heavy investment is necessary for electrification, grid modernization, and networking in these regions. Data privacy and the safety of data are also important points to be considered for the safety of individuals and organizations. Present regulations for e-waste are insufficient and vary widely from one nation to another. For example, it may happen that one company from a developed nation may not insist on recycling in its own country due to the heavy cost of recycling, but the same company may plan the same work in another developing country due to non-stringent recycling laws. So, the place of dismantling of ICT items has changed, but the amount of pollution has not been reduced. So, there is an urgent requirement for global policy and laws. Climate change is a global challenge; hence, coordinated efforts are needed at a global level. Organizations like the United Nations (UN) and the International Telecommunication Union (ITU) work for global ICT sustainability targets.

7.7 Future Trends in Green ICT

We are seeing the ICT revolution. We have seen many generations of communication equipment like 2G, 3G, 4G, and 5G. We have also seen the bursting bubble of the dotcom, and the panic of Y2K! Every time, we have risen and found the way through innovation. Can you list a few innovative ideas for the future?

- 6G technology may knock on our doors any time, which will enable the transportation of data in gigabytes. Super-fast streaming of UHD videos may be possible. Remote operated surgeries may be a new field of medical science, where a doctor from Ahmedabad can operate on a patient on an operating table in New York!
- Better weather forecasting will be possible due to the invention of superfast computers.
- Today, Quantum computers are in the experimental stage, but soon they may replace traditional computers, which will require a much smaller amount of energy.
- AI, IoT, and sensors will penetrate more into our lives with improved technologies to reduce the energy consumption in industries, agriculture, transportation, space technology,

medicine, and our homes.

- Net-zero and circular economy will soon be mandatory provisions in all countries to conserve resources and energy.

-:EXERCISE:-

Section A: Short Answer Questions

1. Define Green ICT.
2. List two goals of green ICT in sustainable development.
3. Mention three environmental impacts of conventional ICT.
4. State the role of energy-efficient computing.
5. Discuss the impacts of e-waste on human health.
6. Describe 'Server virtualisation'.
7. State three advantages of cloud computing because of sustainability.
8. List the advantages of 'green data centres'.
9. List two international initiatives promoting green ICT.
10. State the role of Green ICT in achieving the UN Sustainable Development Goals.

Section B: Long Answer Questions

1. Explain the concept of Green ICT for its contribution to environmental sustainability.
2. Discuss the need for eco-friendly designs of hardware to increase the lifecycle of ICT devices.
3. With examples, explain the role of cloud computing in supporting sustainability.
4. List the strategies to reduce the carbon footprint of ICT devices.
5. Analyse the challenges in adopting Green ICT from the perspective of developing countries like India.
6. Discuss the role of Green Data Centres in energy saving.
7. Explain in detail the significance of e-waste recycling.
8. Suggest a road map for the implementation of Green ICT practices in your college.
9. Assess the role of international collaborations and policy frameworks to foster Green ICT globally.

10. Explain the need for Life Cycle Assessment (LCA) for any gadget.

Section C: Multiple Choice Q and A

1. The ICT stands:
   a) Industrial Control Technology
   b) Information and Communication Technology
   c) Integrated Computing Technology
   d) International Communication Technology
2. Green ICT aims to:
   a) Increase energy consumption
   b) Reduce environmental impact
   c) Expand fossil fuel usage
   d) Promote e-waste dumping
3. ..... is not an ICT device.
   a) Smartphone b) Modern TV
   c) Refrigerator d) Laptop
4. The ICT sector consumes ~___% of global electricity.
   a) 1% b) 4% c) 10% d) 15%
5. By 2030, the ICT sector may require..... energy of total demand..
   a) 5% b) 10% c) 15% d) 20%
6. Power Usage Effectiveness (PUE) measures:
   a) Data centre efficiency
   b) Carbon emissions from vehicles
   c) E-waste recycling rate
   d) Smartphone battery life
7. A PUE value of 1.0 indicates:
   a) All power is used for computing
   b) Inefficiency. Excessive cooling needs
   c) High
   d) Poor IT equipment performance.
8. .... company has achieved a PUE as low as 1.1.
   a) Apple b) Google c) IBM d) Samsung
9. DCiE is:
   a) The inverse of PUE
   b) A cooling system standard
   c) A smartphone battery metric

d) A measure of e-waste generation

10. A typical DCE value is:
    a) 0.5 b) 1.0 c) 1.5 d) 2.0
11. The major ICT environmental impact is:
    a) Noise pollution and deforestation
    b) Greenhouse gas emissions and E-waste
    c) Water pollution and soil erosion
    d) Air pollution and plastic waste
12. A large data centre consumes:
    a) 1-10 MW b) 10-100+ MW c) 500 MW d) 1000 MW
13. A single cellular tower consumes ~:
    a) 1 KW b) 10 kW c) 100 kW d) 1000 kW
14. Approx. Global smartphones:
    a) 1.9 billion b) 2.5 billion c) 6.9 billion d) 10 billion
15. Manufacturing of ……. contributes most of the carbon footprint.
    a) Servers b) Smartphones
    c) Desktop PCs d) Data centre racks
16. E-waste includes:
    a) Discarded plastic bottles
    b) Obsolete electronic devices
    c) Paper waste
    d) Food waste
17. Global e-waste in million tons as of 2022:
    a) 22 b) 42 c) 62 d) 100
18. % of e-waste formally recycled:
    a) 10% b) 25% c) 50% d) 75%
19. 1 ton of mobile phones recovers more ___ than 1 ton of gold ore.
    a) Silver b) Copper c) Gold d) Platinum
20. Largest e-waste by weight:
    a) Laptops b) Smartphones
    c) Large appliances d) Network routers
21. Sustainable design focuses on:
    a) Increasing energy use
    b) Ignoring the carbon footprint
    c) Using recyclable materials
    d) Reducing product lifespan

22. Modularity helps in:
    a) Increasing e-waste
    b) Easy repair/upgrades
    c) Raising costs
    d) Reducing performance
23. Lifecycle Assessment (LCA) identifies:
    a) Cheapest method
    b) High-impact components
    c) Best marketing
    d) Fastest data speeds
24. Green manufacturing involves:
    a) Ignoring recycling
    b) Reducing waste/toxics
    c) Increasing transport emissions
    d) Using fossil fuels
25. Extended Producer Responsibility (EPR) means:
    a) Take-back after use
    b) Landfill disposal
    c) Exporting e-waste
    d) No recycling
26. ICT4S stands for:
    a) ICT for Speed
    b) ICT for Sustainability
    c) ICT for Security
    d) ICT for Servers
27. Smart grids help:
    a) Increase waste
    b) Automate load adjustment
    c) Block renewables
    d) Reduce internet speed
28. Adaptive traffic lights reduce:
    a) Fuel use
    b) E-waste
    c) Data centre efficiency
    d) Internet connectivity
29. EVs rely on ICT for:
    a) Battery management
    b) Slower charging
    c) Reducing recycling
    d) Increasing emissions
30. Agriculture drones help:
    a) Increase water waste
    b) Spray pesticides
    c) Block sunlight
    d) Reduce crop yield
31. Green ICT adoption challenge:
    a) Low-cost
    b) High upfront investment
    c) Nil
    d) No benefits
32. Quantum computers will:
    a) Use more energy
    b) Replace traditional PCs efficiently
    c) Slow processing
    d) Increase e-waste

33. 6G enables:
   a) More emissions b) Remote surgeries
   c) Slower speeds d) Less data transmission
34. The circular economy focuses on:
   a) More e-waste b) Reuse/recycling
   c) Single-use d) Ignoring sustainability
35. Net-zero means:
   a) No internet
   b) Balance emissions/removal
   c) Only fossils
   d) No recycling
36. The ICT carbon footprint depends on:
   a) Electricity mix b) Internet speed
   c) Data centre colours d) Smartphone count
37. ..... is not found in smartphone batteries:
   a) Lithium b) Nickel c) Lead d) Cobalt
38. Ideal PUE:
   a) 0.5 b) 1.0 c) 1.5 d) 2.0
39. Global ICT sustainability body:
   a) WHO b) ITU c) NASA d) FIFA
40. AI/IoT help:
   a) Increase waste b) Block smart grids
   c) Optimise energy d) Reduce efficiency

41. **True or False:** Green ICT only focuses on reducing e-waste.
42. **True or False:** Data centres consume less power than smartphones.
43. **True or False:** Smartphone manufacturing generates a larger carbon footprint than its lifetime use.
44. **True or False:** E-waste is a threat to human health.
45. **True or False:** E-waste is increasing at an alarming rate.
46. The term __________ means discarded electronic devices.
47. ___________ is a metric to measure data centre efficiency.
48. ___________ involves manufacturers taking back products after use.
49. -------------- helps in spraying pesticides in agriculture.
50. ___________ computers may reduce energy consumption in the future.

Answer Keys

1. B, 2.B, 3.C, 4.B, 5.C, 6.A, 7.A, 8.B, 9.A, 10.A,
11.B, 12. B, 13. B, 14.C, 15.B, 16.B, 17.C, 18.B, 19.C, 20.C,
21. C, 22. B, 23. B, 24. B, 25. A, 26. B, 27. B, 28.A, 29. A, 30. B,
31.B, 32.B, 33.B, 34.B, 35.B, 36.A, 37.C, 38.B, 39.B, 40.C,
41.F, 42.F, 43.T, 44.T, 45.T, 46. E-waste, 47.PUE, 48.EPR,
49.Drones, 50.Quantum.

वायोर्जलस्य तेजश्च, पृथिव्याश्च नियोजने।
गृहं स्थाप्यं यथाशास्त्रं सुखं जीवेत् गृहाश्रयी॥

(Inspired by the Ancient Vaastu Shastra book Mayamata and Manasara)

"Let the building be constructed in alignment with Air, Water, Fire (that is Energy), and Earth (that is Material) so that the residents can live peacefully and happily."

This shloka emphasises the modern principle of sustainable architecture:

- Air: Natural Ventilation
- Water: To express water conservation and harvesting
- Energy (Tejas): To harness the solar energy.
- Earth: To use materials judiciously

It implies that residents can live happily and peacefully only when construction respects the natural forces: **an ancient call for climate responsibility and sustainability.**

Chapter 8

GREEN BUILDING AND URBAN PLANNING

Learning Outcomes:

After reading this chapter, the learner will be able to:

1. Understand the key principles of green building and certification (e.g., LEED, BREEM, IGBC).
2. Discuss the passive solar architecture and net-zero buildings.
3. Suggest sustainable urban strategies to reduce energy consumption.
4. Explore the role of automation to conserve energy.
5. Analyse the need for energy-efficient equipment in domestic and industrial sectors (e.g., LED)

8.1 Green Buildings

As mentioned in the previous page, through a Sanskrit text mentioned in the Ancient Vaastu Shastra book Mayamata, humans should live amicably with their surroundings and the ecosystem. Our ancestors were able to understand the importance of aligning with nature and sustainability. In the same line, the modern concept of Green Building implies the construction of a building in such a way as to minimize the impacts on the environment while maintaining the comfort and well-being of its occupants. It emphasises the thriftiness and conservation of water, energy, and other natural resources while reducing pollution and waste and extending the building's life. Key features of green buildings include energy efficiency by maximizing the use of renewable energy (like natural daylight, solar, and

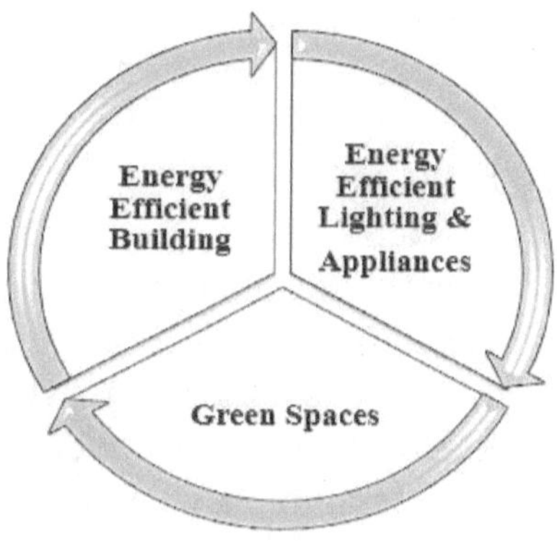

Figure 8.1 Pillars of Green Buildings

wind); water conservation (by rainwater harvesting, water recycling); use of locally available, recycled, low-carbon-footprint, non-poisonous materials; waste reduction; and sustainability through green roofs, gardens, permeable pavements, for minimum disruptions to the environment.

> ***Thinking Corner**: Green buildings often win awards, but can the poor or middle-class people like you and me afford to live in them? Whether we call it development or sustainability, it must be inclusive. Designs must be accessible, not just aspirational.*
>
> *What's your opinion?*

Briefly, we may say, Green Building refers to the concept of the comprehensive approach of designing a building in such a way as to be environmentally responsible, economically viable, and socially supportive, to utilise the maximum possible natural renewable resources at their best.

8.2 Green Architecture Principles

Mass industrialisation has led to rapid urbanisation, requiring heavy construction. Multi-story buildings have increased the population density as hotspots in urban areas. Nearly 35% of carbon dioxide is emitted directly or indirectly due to these buildings, consuming a great part of natural resources. In response to climate change and global warming, the concept of green building has been codified through certain certification systems. Table 8.1 shows a few green building standards followed by a few leading countries.

Key Principles

- Energy Efficiency: Buildings should be designed with proper insulation (to retain heat in colder countries), proper ventilation for cooling, and passive solar design to reduce solar radiation on walls. It should be equipped with renewable energy sources like wind, PV solar for electricity generation, and geothermal heating to keep the building warm.

Sr. No.	Country	Standard	Full Form
1	USA	LEED	Leadership in Energy and Environmental Design
2	UK	BREEM	Building Research Establishment Environmental Assessment Method
3	India	IGBC	Indian Green Building Council
4	India	GRIHA	Green Rating for Integrated Habitat Assessment

Table 8.1 Green Building Standards

- Water Conservation: Rainwater harvesting, and water recycling systems should be deployed to store and conserve water. Reducing the water use by using low-flow taps and landscaping with drought-resistant grass and plants.
- Sustainable Materials: Materials should be renewable, locally available, non-toxic, and have a minimum carbon footprint. If possible, recycled or rapidly renewable.
- Waste Reduction: To reduce the construction and demolition waste. To recycle the maximum amount of waste with or without processing. To compost the material if possible.
- Site Sustainability: To reduce the ecological impact by minimising the use of land and resources, and protecting biodiversity.

To certify and implement the above concepts, many organisations, as mentioned in the above table, exist. Let us understand it in detail:

8.2.1 IGBC (Indian Green Building Council)

It was launched in 2001 by the Confederation of Indian Industries (CII) and is affiliated with the US Green Building Council (USGBC) and LEED. It is a voluntary rating system designed to suit the Indian climate and perspective. It offers ratings for IGBC Green New buildings, IGBC Green Homes, IGBC Existing Buildings, IGBC Green Factories, IGBC Green Cities, Townships, and

Schools. It focuses on areas like sustainable site planning, water and energy conservation, optimisation of materials, indoor environment quality, and innovative passive designs. It issues certificates in the categories of Silver, Gold, and Platinum. It has been widely accepted in the Indian industrial panorama due to its flexibility and alignment with international organisations like LEED. Due to high green cover, zero waste incineration, energy efficiency, and water harvesting, IIT Madras, Chennai, has been awarded the IGBC Green Campus Platinum certificate in 2023. Similarly, the National Institute of Construction Management and Research (NICMAR), Hyderabad, and Ahmedabad University, Gujarat, have also been awarded IGBC Platinum certificates for water reclamation, rooftop solar plants, and biodiversity enhancements. In the industrial and corporate sector, Godrej Plant 13 Annexe, Mumbai, has been awarded the IGBC Platinum certificate [40]. Normal buildings may consume nearly 30% of India's electricity, which can be reduced by 50% through sustainable architecture (source BEE India) [41],[42]. Readers can refer to NBC 2016 and the Energy Conservation Building Code (ECBC) for more reference and standards needed in the professional sector.

8.2.2 GRIHA (Green Rating for Integrated Habitat Assessment)

It has been developed by The Energy and Resource Institute (TERI) and is endorsed by the Ministry of New and Renewable Energy (MNRE), Government of India. It has also been tailored to suit the demands of the Indian climate, building codes, and cultural and social contexts. It may be used for residential, commercial, or institutional new buildings, renovated buildings, and cities. It focuses on proper site planning, energy efficiency, integration of renewable energy sources, solid waste management, socio-economic well-being, water management, and indoor environment quality (IEQ). It emphasises passive designs and low-cost sustainable practices. It is aligned with India's National Mission of Climate Change. It awards the certificates in a star system based on points out of 100 as below:

GRIHA Certification Level	Out of 100 Points
1 Star	50-60
2 Star	61-70
3 Star	71-80
4 Star	81-90
5 Star	91-100

Table 8.2 GRIHA Certification Level

Now, let us compare these two certification systems:

Sr. No.	Feature	IGBC	GRIHA
	Full Form	Indian Green Building Council	Green Rating for Integrated Habitat Assessment
1	International Alignment	LEED	Indigenous. Tailored to Indian codes and climate
2	Rating Style	Points + Certification Platinum, Gold, Silver	Points out of 100 + Star rating (1 to 5)
3	Developed by	CII (with USGBC)	TERI (with MNRE)
4	Emphasis	Energy, water, and material	Passive design, climate adaptation
5	Popularity	Urban housing, Commercial, and corporate	Government and institutional buildings
6	Tools	Check list	Comprehensive
7	Endorsement	No official endorsement or affiliation	Endorsed by MNRE, Government of India
	Cost of certification	Higher, depending on the consultant	Moderate, cost-effective, and more transparent
	Mandate	Not mandatory	Mandatory for most projects
	Preferred for	IT parks, malls, offices, urban townships, and industries	Government offices, schools, universities, etc

Table 8.3 Comparison between IGBC and GRIHA standards

8.2.3 LEED Certification

LEED (Leadership in Energy and Environmental Design) was developed by the US Green Building Council (USGBC) in 1998.

It issues ratings for the diverse types of buildings, like schools, offices, commercial buildings, and hospitals. It is one of the most widely used certification systems in the world due to its comprehensive rating system. It evaluates the building under different criteria:

1. Sustainable Site
2. Water Efficiency
3. Energy and Atmosphere
4. Materials and Resources
5. Indoor Environmental Quality
6. Innovation
7. Regional Priority
8. Integrative Process

LEED Certification Levels:
Buildings under consideration are evaluated by the above criteria and are awarded the following certifications based on points received out of 100:

- Certified: 40-49 points
- Silver: 50-59 points
- Gold: 60-79 points
- Platinum: 80+ points

Advantages: Less water and energy are consumed in LEED-certified buildings. Thus, lower operational costs, reduced utility bills, and growing consumer preference for green buildings. Occupants are prone to having a healthier life due to better IEQ. Tax benefits from governments to green buildings are also beneficial. LEED is also a globally acceptable parameter.

8.2.4 BREEM Certification

BREEM is a Building Research Establishment Environmental Assessment Method. It was the first sustainable method in the world, which was launched in 1999 in the United Kingdom. It is one of the most popular certification methods in Europe. BREEM is flexible and evidence-based to assess the cost of life cycle and ecological assessment. BREEM evaluates any building based on the following criteria:

1. Management
2. Health and Well-being
3. Energy
4. Transport
5. Water
6. Materials
7. Waste
8. Land Use and Ecology
9. Pollution
10. Innovation

BREEM certifies the building as pass, good, very good, excellent, and outstanding as per the following points. This point system has been tailored to suit various regions.

- Pass >=30%
- Good >=45%
- Very Good >=55%
- Excellent >= 70%
- Outstanding >=85%

Role of Green Certifications in Sustainable Development:

Sr. No.	Sustainable Development Goals	Remark
1	SDG 7	Affordable and Clean Energy
2	SDG 11	Sustainable Cities and Communities
3	SDG 12	Responsible Construction and Production
4	SDG 13	Climate Action

Table 8.5 SGS and Green Certifications

Comparison between LEED and BREEM

Sr. No.	Criterion	LEED	BREEM
	Origin	USA (1998)	UK (1990)
	Certification Authority	US Green Building Council	Building Research Establishment, U.K.
	Spread	Global	Europe
	Focus	Energy, Water, Site, Materials, Innovation	Management, Ecology, Pollution, Lifecycle
	Customization	Limited	Region-specific adaptation
	Process	Transparent, rigid point system	Flexible weightage
	Strengths	Standard framework, global brand	Adaptable, Ecological emphasis
	Others	Design stage compliance	Includes post-construction evaluation
	Global Reach	Used in more than 180 countries	Dominant in the UK and Europe
	Categories	9	10
	Scoring System	Fixed point system	Weightage based
	Ecology and Biodiversity	Moderate	Strong

Table 8.4 Comparison between LEED and BREEM

> ***Thinking Corner**: Tradition Meets Innovation*
>
> *Our ancestors built cool homes without ACs, shaded courtyards, and mind-blowing living space to easily live without electricity. Are we forgetting indigenous wisdom in the sheep race to modernize sustainable buildings? Are we really progressing?*

8.3 Passive Building Designs and Energy-Efficient Materials

Passive building refers to the architectural design that focuses on the natural environment to maintain comfort for the residents with minimum energy consumption. It also focuses on using energy-efficient devices and materials. It also emphasises building's structure, material, and design to use natural energy forms like sunlight, wind, and thermal mass to cut energy consumption at an optimum level. Passive design principles are not new. Our ancestors have been using them for centuries. The main principles involved in passive building designs are as follows:

1. Proper Orientation relative to the sun's path to maximise the natural sunlight and heat during winter.
2. Thermal Mass, like concrete and brick, helps store heat to stabilize the indoor temperatures.
3. Particularly in cold countries, proper sealing and high-quality insulation help in reducing heat loss.
4. Placement of doors and windows should be such as to maximise the natural air circulation and ventilation.
5. Proper design to reduce the use of artificial lighting and maximise the natural daylight.
6. To prevent unwanted solar heat overhangs, curtains, louvers, and windows should be properly placed for shading.

Energy Efficient Materials:

The selection of proper material is especially important to enhance the building's energy performance. They contribute to passive performance for energy reduction and increase ecological sustainability. The following are a few materials worth discussing:

- Insulation Materials: Insulation reduces the transfer of heat from walls, roofs, floors, and windows. Fibreglass is the most cost-effective and common insulation material, followed by mineral wood due to its fire-resistant and sound-absorbing properties. Cellulose is made from recycled paper; it is eco-friendly and effective, too. Spray form has high sealing properties but is more expensive.
- Windows and Glazing: Maximum heat gain or heat loss occurs from windows; hence, they need more attention. So, low-emissivity glass may be used in windows, and thermal black films may help to reduce the heat flow.
- Thermal Mass Materials: These materials slowly absorb and release energy. Concrete and brick are excellent thermal mass, but have high embodied energy. They have lower carbon footprints and are good for dry climates.
- Roof: Roofs should be made from such materials that reflect maximum light, keeping the building cool. For this, white membranes, cool metal roofs, ceramic coatings, green cover on roofs, and roof plantation may help in keeping the roof cool.
- Interior: Bamboo flooring, recycled tiles, carpets, low VOC paints, etc, may help the building remain cool and renewable.

To enhance the performance of green buildings, passive designs may be integrated with smart technologies. For example, automated blinds to adjust the sunlight intensity, Occupancy sensors to reduce lighting depending upon the number of people in the building, and smart thermostats to adjust cooling based on occupancy, following green building standards may help improve performance. Bullitt, Seattle, USA, is a commercial designed to be a net energy zero and net water zero building, having deep overhangs, composting toilets, and rainwater harvesting. Indira Paryavaran Bhavan, New Delhi, India, is the first net-zero energy building built by the Government of India. It achieved a 70% reduction in energy consumption. Bahnstadt district, Heidenberg, Germany, has also reduced energy consumption by over 50%, combining urban planning and a passive housing system. Use of passive building design and energy-efficient materials not only helps reduce energy consumption but also reduces operational costs, reduces carbon footprint, enhances residents' health and comfort, while uplifting the asset value. However, high initial costs, complexity of designs, and regulatory gaps are the few challenges to cope with.

> ***Thinking Corner**: Concrete or Conscience?*
> *Do you know, cement is the second most consumed material on Earth after water! So naturally, a major emitter of CO_2!*
> *Isn't it possible to reimagine and redesign our cities with natural materials like bamboo, clay, and recycled materials? Now think of the "Gram Swaraj" Dream of Mahatma Gandhiji.*

8.4 Green Cover

In recent decades, urbanization has accelerated globally, leading to environmental degradation. Concrete jungles are eating the forests slowly, which has resulted in urban heat islands. To mitigate these challenges, gardens, urban green spaces, and rooftop gardens may help multidimensionally. These green spaces in the form of community gardens, parks, urban forests, lane gardens, rooftop gardens, balcony plantations, and green corridors are important components for social, economic, and ecological benefits. The following are a few advantages:

- Air pollutants are filtered by trees and plants. They absorb carbon dioxide and release oxygen.
- They provide a cool environment through transpiration and shade.
- They also help in stormwater management by absorbing rainwater.
- They counteract the heat island effect by temperature moderation, reducing the temperature of the surface and air.
- They also help in the reduction of noise.
- They are an important part of the biodiversity.
- They also offer social and psychological benefits.

- They directly help in mitigating the effects of climate change.
- They can help increase fresh urban food availability.
- They provide a place to play for kids and a place for community and social engagement.

Rooftop Gardens:

Rooftop gardens are the green spaces provided in the buildings on the terrace. They may be classified as follows. Extensive Green Roofs: These are plantations on the top of a building or terrace, depending upon the design of the building. They require less maintenance and are lightweight. They are planted in a shallow growing medium of up to six inches. These are suitable for stormwater absorption and insulation.

Intensive Green Roofs: They are more suitable for a deeper growing medium of more than six inches. They may include a variety of plants, shrubs, and small trees. They are heavier and require more maintenance. They may serve the purpose of recreational parks or community spaces.

Semi-intensive Green Roofs: They combine the good features of both systems above, as per the suitability. They are used for aesthetic appeal to decorate the building with partial human access. They do have limited vegetation.

The design of a rooftop garden involves careful planning and site assessment. Structural integrity, sunlight, and wind exposure should also be considered. Safety measures should also be kept in mind, given needed maintenance, and human access. The green roof system includes a vegetation layer, a growing medium, a filter fabric, a drainage layer, and a waterproof membrane. Root barrier and thermal insulation. Plant species should be selected based on the suitability of local climatic conditions and irrigation requirements. Seeing the importance of greenery, rooftop gardens have been integrated into urban planning and green building codes.

> ***Activity:*** *Education at the cost of Earth?*
> *Calculate the approximate total number of pages (and number of trees to manufacture those notebooks/textbooks) you might have used since your birth till you complete your education.*
> *Note: Global average: 8,000 pages, equivalent to 4.8 trees.*
> *How many trees have you planted so far?*

Case Studies:
Singapore is known as a Garden City due to its urban green cover of over two hundred hectares of roof gardens and vertical gardens. Marina One is a prominent example of a rooftop garden designed with biodiversity corridors. The Government of Singapore has framed a few regulations and motivated vertical and rooftop greenery through the Skyrise Greenery Incentive Scheme. New York City also offers a substantial amount of tax benefits for promoting green cover in urban areas and buildings. Brooklyn Grange is one of the largest rooftop soil farms located in New York City, which produces more than 25,000 kg of vegetables per annum. On the other hand, Copenhagen has mandated green roofs for all new buildings. Copenhagen aims to become a carbon-neutral city by 2025. Rooftop and vertical gardens are playing a significant role in this mission.

Challenges:
Cost and structural load for old buildings can be very costly and complex to install green cover. Proper maintenance is a big challenge, as without regular maintenance, plants will die, and the system can fail. Irrigation is also a big challenge in dry climate areas. Many financial incentives are required to encourage people to opt for roof gardens and vertical gardens.

8.5 Energy-Efficient Appliances

The concept of green building expects the reduction of energy consumption in every viable way. Energy-efficient appliances and lighting systems play a key role in decreasing the requirement for electrical energy. Thus, they can reduce the greenhouse gas emissions and operating costs for a house, office, or industry. Energy-efficient appliances and monitoring systems have become an integral part of green buildings.

THE ENERGY STAR rating system is a widely used and well-recognised system to show the comparative consumption of a particular appliance. It was launched in 1992 by the US Environmental Protection Agency (EPA). It is an easy and convenient system for the identification of energy-efficient products. This system covers a wide variety of consumer products like washing machines, dishwashers, refrigerators, lighting, TV, AC, ovens, and many other electronic products. Star-certified products assure consumers of energy saving without compromising performance. In India, the Bureau of Energy Efficiency (BEE) Star Rating system uses a 1-to-5-star label, where any appliance with a label '5 stars' represents the most energy efficient and '1 star' represents the lowest energy efficiency. Even a 1-star appliance will also be saving more energy than a conventional appliance.

The EU energy label ranks appliances from A+++, as most efficient, to G, as least energy efficient. On the other hand, customers are informed through possible annual energy consumption and cost comparison by the Energy Guide (USA). The top runner program of Japan sets the next standard for energy efficiency based on the performance of present appliances. Thus, we see that different regions have different standard systems to guide their people in the wise selection of energy-efficient appliances.

Lighting:

The invention of the light bulb was a cumulative process involving several inventors like Humphry Davy, Warren de La Rue, and Joseph Swan. But the first reliable, long-lasting, and commercially viable incandescent lamp was invented by Thomas Edison in 1879. Lighting is important to light our homes during evenings and nights, in industries, streetlights, and in many other applications. Lighting has brought a revolution in human history. Initial lighting sources like incandescent lamps were consuming more energy. Newer lighting sources like tube lights and compact fluorescent lamp (CFL) have improved the energy efficiency significantly, but the latest Light Emitting Diode (LED) technology was the biggest breakthrough in this field, which consumes a small amount of energy. LEDs can also provide distinct colours. Around 20% of the total energy is consumed in the lighting sector.

Sr No.	Feature	Incandescent Bulb	CFL	LED
1	Lifespan	1000 hours	8000 hours	50000 hours
2	Heat Emission	High	Low	Minimum
3	Environmental Impact	Tungsten is used	Mercury is used	Non-toxic, recyclable
4	Energy Consumption	High	Moderate	Very Low
5	Colour Options	Very Limited	Few	Wide range
6	Efficiency	10-15 lm/W	35-50 lm/W	80-125 lm/W
7	Size	Moderate	Moderate	Very Small

Table 8.6 Comparison of Various Lighting Sources

Table 8.7 represents the luminous efficacy for various light sources, which is the amount of light produced in lumens per one watt of energy consumed. The table clearly shows the reason LEDs are dominating the market.

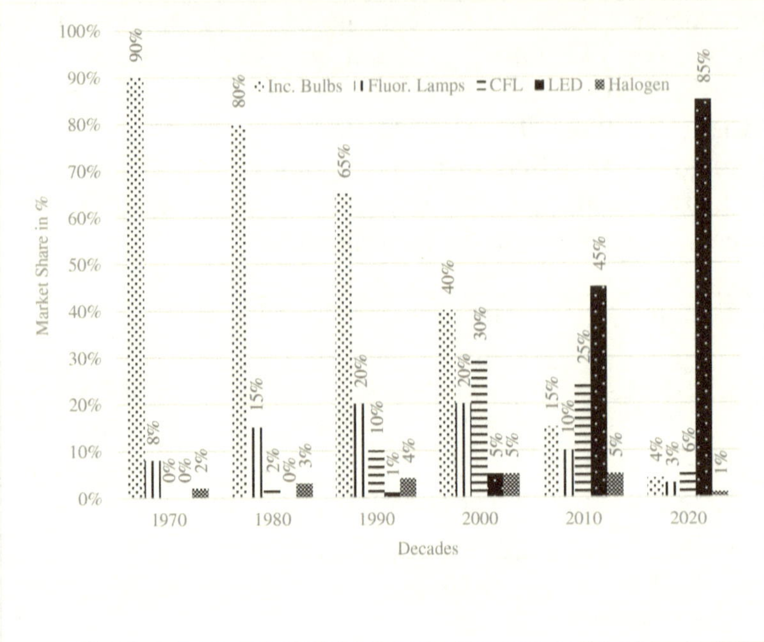

Figure 8.2 Declining Market Share of Lamps Vs LEDs in the last 50 years (Note: Values are indicative based on historical data patterns from various sources like IEA, US DOE, and lighting market reports as of 2024)

Advantages of LED Lighting

- LEDs have extremely high efficiency, resulting in extremely low electricity bills.
- LEDs have longer life, hence a minimum replacement frequency and costs.
- LEDs have the characteristic of Instant-ON with full brightness.
- LEDs are dimmable and programmable; hence, they are compatible with smart systems.
- No hazardous substances are used in the manufacturing of LEDs.

Sr. No.	Light Source	Luminous Efficacy In Lumens per Watt	Remark
1	Incandescent Lamp	10-18	Very inefficient, major energy is lost as heat
2	Halogen Bulb	16-25	Poor efficiency
3	Compact Fluorescent Lamp	45-70	More efficient, longer lifespan than incandescent lamps
4	Fluorescent Tube (T5/T8)	60-100	Popular in commercial buildings, higher efficiency
5	Metal Halide Lamp	65-115	Use in industrial and sports stadiums
6	High-Pressure Sodium Lamp (HPS)	70-140	Widely used in streetlights. Good efficiency but poor colour rendering
7	LED (Standard)	90-120	The most popular choice in today's market
8	LED (High efficiency)	150-230	Costly, used in commercial, industrial, and smart lighting
9	Organic LED (OLED)	20-60	Currently, it is a less efficient emerging field, popular in displays rather than lighting.

Table 8.7 Luminous efficacy for various light sources

Due to these types of benefits, LEDs are universally used as a primary lighting source. They are not just used in homes but are also used in offices, for streetlighting, in commercial buildings, and many other places, contributing to reduced power demand and GHG emissions.

Smart Appliances and Automation Systems:

Recent decades have witnessed a revolutionary boom in the use of the internet and the Internet of Things (IoT) devices. With the help of the internet and IoT devices, it is now possible to have local as well as remote control of many electrical appliances. This has resulted in improved energy efficiency, energy conservation, and responsiveness.

Smart Appliances integrate sensors, onboard chips for local /remote control, and automation to optimize energy use. Smart washing machines can schedule the operation of washing clothes at non-peak hours to take advantage of lower tariff rates. It can also automatically adjust the water and electricity needs based on the number of clothes to be washed. A smart refrigerator can automatically adjust the temperature based on door openings, usage patterns, and the volume of items placed in the refrigerator. A smart thermostat can regulate the heating and or cooling as per the number of occupants. These systems help in the reduction of electricity bills and help environmental sustainability.

Building Automation Systems (BAS) are used in commercial buildings to control lighting, energy use, cooling, and heating. This used to be the centralized system pivoted on the web interface or mobile application. They are mostly integrated with predictive analysis to conserve energy using an energy forecast to prevent overuse. It is desirable to get power from renewable energy sources for the BAS to make it more sustainable. It is possible to save up to 30% of energy in large commercial buildings using intelligent controls and BAS. They can also provide real-time probable faults and their alerts for maintenance needs in advance.

Sensors and timers play a significant role in IoT and BAS. Sensors can detect movement and automatically turn ON or OFF lights, while timers can turn ON or OFF any appliance based on preset times. Sensors are useful in restrooms, public halls, and storage areas. Timers are useful in streetlights, signage, and commercial appliances. Environmental sensors measure temperature, humidity, CO_2, and the level of occupancy for auto climate control. BAS works on real-time data provided by these sensors.

Activity: Observe various electrical items in your home. How many of them are energy efficient? Calculate the cost of replacing the old devices, like lamps, with LEDs. Calculate the payback period for the same.
Motivate your parents and neighbours to use BEE-certified appliances like washing machines, refrigerators, and TVs with star ratings.

8.6 Energy Efficiency in Different Sectors

- Residential Sector: All appliances like ACs and washing machines should be equipped with inverter technology to conserve power. Solar-powered appliances like Solar water heaters and solar cookers should be promoted.
- Commercial Sector: Heavy power-consuming appliances like commercial refrigerators should be equipped with high-efficiency compressors. The BAS system should be invariably integrated to conserve maximum energy. A motion-controlled lighting system should be used. Energy-efficient appliances should be used. LED display panels should be used rather than conventional CRTs.
- Industrial Sector: High-efficiency motors should be used. Wherever possible, variable frequency drives should be used for the motors having long running durations. The load shedding strategy should be planned to reduce the maximum demand. An energy audit should be compulsory every year.

-:EXERCISE:-

Section A: Short Answer Questions

1. Define Green Building.
2. List any two principles of Green Architecture.
3. State the full forms of LEED and BREEM.
4. Mention the benefits of locally sourced construction materials.
5. Define Passive Solar Design.

6. List three technologies to increase energy efficiency in green buildings.
7. Define net-zero building.
8. Discuss the role of green roofs in sustainability.
9. List two water conservation techniques in green buildings.
10. State the role of sensors in green buildings.

Section B: Long Answer Questions

1. Explain the objectives of Green Building. Mention how it differs from conventional buildings.
2. Discuss the main components of sustainable building.
3. Compare the criteria for LEED and GREEM certification systems.
4. Elaborate on the role of local construction materials in reducing the carbon footprint of a building.
5. Analyse the role of a solar PV rooftop system in reducing the carbon footprint of a building.
6. Discuss the role of passive design to enhance the efficiency of buildings.
7. Discuss the social, economic, and environmental benefits of green buildings in urban planning.
8. Assess the role of wastewater recycling in sustainability.
9. Propose a roadmap to implement the green building policy in Indian Educational Institutions.
10. Discuss the role of green buildings in mitigating climate change.

Section C: Multiple Choice Q and A

1. The primary goal of green building is:
   A) Luxury design
   B) Environmental sustainability
   C) Maximum profit
   D) Cheap
2. The .... ancient text first mentioned eco-friendly construction principles.
   A) Arthashastra B) Mayamata
   C) Vedas D) Geeta

3. .... is NOT a key feature of green buildings.
   A) Energy-efficient lighting B) Solar panels
   C) Single-pane windows D) Rainwater harvesting

4. LEED certification is developed by:
   A) US Green Building Council
   B) World Green Building Council
   C) Building Research Establishment
   D) Indian Green Building Council

5. The IGBC stands for:
   A) International Green Building Code
   B) Indian Green Building Council
   C) Indian Green Building Certification
   D) Integrated Green Building Concept

6. The star-based certification system is used by:
   A) LEED B) GRIHA C) BREEM D) IGBC

7. The LEED certification assigned the highest level is:
   A) Gold B) Silver C) Platinum D) Diamond

8. The first green building certification system is:
   A) LEED B) BREEM C) IGBC D) GRIHA

9. The .... material is best for thermal mass in buildings.
   A) Plastic B) Wood C) Concrete D) Glass

10. The main purpose of low-emissivity glass is:
    A) Block all sunlight B) Reduce heat transfer
    C) Increase heat transfer D) Increase UV radiation

11. The... roofing material is LEAST energy efficient.
    A) Green roof B) Black asphalt
    C) White membrane D) Cool metal

12. LED lifespan is .... than incandescent bulbs.
    A) Can't say B) Similar C) 5x longer D) 50x longer

13. The ...... lighting source has the highest luminous efficacy.
A) Incandescent bulb B) CFL
C) LED D) Halogen

14. The CFL bulbs have the main disadvantage of:
A) High-cost B) Contains mercury
C) Short lifespan D) Low brightness

15. The BEE Star Rating indicates:
A) Luxury level B) Energy efficiency
C) Cost-effectiveness D) Durability

16. Which country introduced the Top Runner Program?
A) USA B) Japan C) India D) Germany

17. The primary benefit of smart thermostats is:
A) Increase energy use B) Reduce energy waste
C) Decrease ventilation D) Raise temperatures

18. The Main benefit of rooftop gardens is"
A) Aesthetic B) Stormwater control
C) More energy use D) Less biodiversity

19. .... City mandates green roofs for new buildings?
A) Singapore B) New York
C) Copenhagen D) Delhi

20. The rooftop gardens have a prominent challenge of:
A) High-cost B) Structural load
C) High maintenance D) Too much rain

21. .... is NOT a green roof type.
A) Extensive B) Semi-intensive
C) Intensive D) Synthetic

22. The main purpose of urban green spaces is:
A) Decrease biodiversity B) Reduce carbon
C) Raise temperatures D) Increase pollution

23. IoT means:
A) Internet of Things B) Integrated Office Tools
C) Institute of Technology D) International Operations Team

24. The ……….. system optimizes commercial energy use?
A) HVAC B) BAS
C) Solar panels D) LED lights

25. The occupancy sensors….
A) Waste energy B) Automate lighting
C) Block ventilation D) Raise temperature

26. Example of a smart appliance:
A) Incandescent bulb B) Manual thermostat
C) Modern refrigerator D) Toaster

27. Estimated energy savings from BAS?
A) 5% B) 15% C) 30% D) 50%

28. The ……. technology is used in efficient ACs.
A) Inverter B) Fixed speed
C) High-power D) Manual

29. The purpose of VFDs in industries is:
A) Waste energy B) Optimize motors
C) Decrease demand D) Reduce lifespan

30. Best lighting for streetlights is:
A) Incandescent B) Halogen
C) LED D) CFL

True or False?

31. LEED certification began in the UK.

32. Green roofs help reduce the urban heat island effect.

33. Concrete has no place in green buildings.

34. IGBC is mandatory for all Indian buildings.

35. Bamboo may be considered a sustainable building material.

36. The LEDs are less efficient than CFL bulbs.

37. Passive design relies on mechanical systems.

38. GRIHA is endorsed by MNRE.

39. Black roofs are ideal for hot climates.

40. Smart appliances increase energy consumption.

41. The Indian rating system developed by TERI is ______.

42. ______ glass minimizes heat transfer.

43. ______ panels convert sunlight to electricity.

44. ______ mass materials stabilize indoor temperatures.

45. Proper ______ reduces the need for AC systems.

46. ______ reduces heat transfer through walls.

47. The ______ footprint is reduced by green buildings

48. From natural forest, ______ is a fast-growing, sustainable material.

49. ______ enables smart appliance connectivity.

50. ______ harvesting conserves water resources.

Answer Keys:

1B, 2B, 3C, 4A, 5B, 6B, 7C, 8B, 9C, 10B, 11B, 12D, 13C, 14B, 15B, 16B, 17B, 18B, 19C, 20B, 21D, 22B, 23A, 24B, 25B, 26C, 27C, 28A, 29B, 30C

True/False

31F, 32T, 33F, 34F, 35T, 36F, 37F, 38T, 39F, 40F

Fill-in-Blanks

41GRIHA, 42Low-emissivity, 43Photovoltaic, 44Thermal, 45Ventilation, 46Insulation,47Carbon, 48Bamboo, 49IoT, 50Rainwater

कृषिं च वा एते प्रजाः स्वधामनु
तस्यां हि धान्यं बहुधा निहितम्।
तस्मात् कृषिं भुञ्जते अन्नदानं
तस्यां हि सर्वं प्रजया संविविष्टम्॥

(Rigveda 10.117.6)

Agriculture is the foundation for prosperity and life. All living beings are nourished and sustained through farming."

Chapter No. 9

SUSTAINABLE AGRICULTURE

Learning Outcomes:

After reading this chapter, the learner will be able to:

1. Define sustainable agriculture to meet food security.
2. Compare different types of farming techniques (Traditional, Organic, Permaculture)
3. Explore the modern techniques of farming, like vertical farming and hydroponics.
4. Discuss the importance of sustainable agriculture for the environment and human health.
5. Discuss the role of smart farming in mitigating climate change.

9.1 Introduction

Our population is growing at an amazingly fast rate. As per the estimation released by the United Nations, the world population exceeded eight billion just before the year 2022. You will be surprised to know that we, humans, took almost 3,00,000 years to reach the 1 billion population level and just 218 years to reach 8 billion! The population growth rate is continuously declining, yet we are adding 140 million people every year [43]. The following table 9.1 and its funnel-shaped figure 9.1 help us better understand how fast we are doubling.

Sr. No.	Years to Double the Global Population (From AD 1700)	Population in billions
1	544	0.375 to 0.75
2	304	0.5 to 1
3	166	0.75 to 1.5
4	123	1.0 to 2.0
5	79	1.5 to 3.0
6	47	2.0 to 4.0
7	39	3 to 6
8	40	6 to 8

Table 9.1 Doubling Trend of Global Human Population.

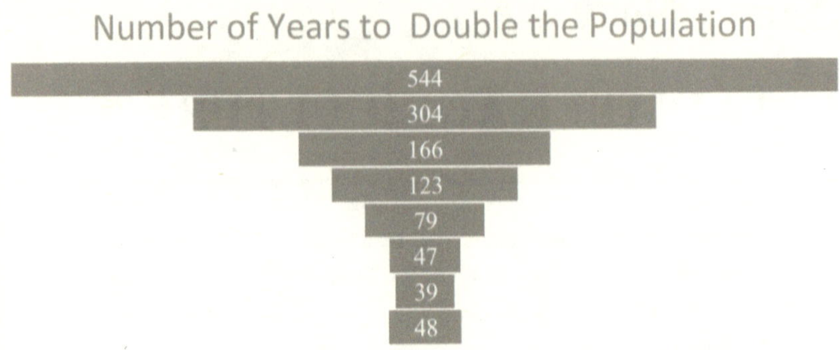

Figure 9.1 Number of years required to double the human population [43].

Now, the fundamental issue that arises in our mind is that of grains to feed this big population and land to accommodate them. We shall focus our discussion on food requirements. In the last few decades, advancements in agricultural technology have increased the yield manyfold. The following graph shows the comparative analysis of food grain production and population growth. From the graph, it is truly clear that we have been able to fulfil the needs of the growing population so far, but now, due to scarcity of land and that's too fertile land, decreasing water levels have raised serious doubts about future food production. It is expected that our global population will become 9.7 billion by 2050, but agricultural production might not be able to keep up with the needs of the population due to the impacts of climate change, droughts, soil degradation, and biodiversity loss. So far, we have been able to meet the demand for food and other agricultural products with the help of chemical fertilizers, improved crops, and soil upgradation, but now it has reached a saturation level. Net agricultural land keeps decreasing to fulfil the growing needs of the increased population, to construct houses and industries, and to fulfil other needs. Further, 33% of the world's arable land has decreased in the last 40 years due to modern farming, which uses synthetic fertilizers, excessive water, and pesticides. So, it is an urgent need of the hour to address the problem of sustainable agriculture.

Sustainable agriculture refers to the production of food to meet the demands of present generations without compromising the needs of future generations. It has very few negative environmental impacts. It can be achieved with the help of technological innovations and ecological stewardship to ensure food safety for all. This chapter has been divided into three major transformative pillars of sustainable agriculture: (a)organic farming and permaculture, (b) AI-based smart farming, and (c) Vertical farming and hydroponics.

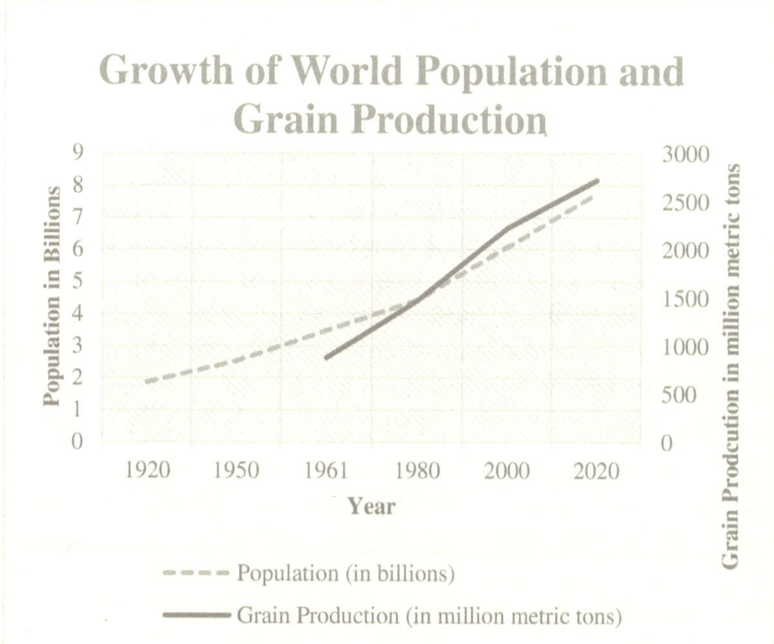

Figure 9.2 Growth of population and Grain Production.

9.2 Organic Farming

Organic farming is an agricultural system that is based on the traditional way of farming, which does not involve the use of synthetic fertilizers, pesticides, or genetically modified organisms. It aims to maintain the soil health and the Agri-ecosystem. It relies on natural products to boost the yield, like biological pest control, compost fertilizer, and green manure. It follows practices like crop rotation to maintain the sustainability of soil health. Thus, it aims to increase biodiversity and reduce the environmental impact of agriculture.

Organic farming has been practiced for centuries; it is not a new or modern invention. It was practiced by our ancestors and Indigenous communities across the globe. The Terra Preta, a nutrient-rich soil, is known for its high fertility. It was created by Amazonian people more than 2000 years ago. It has high carbon content and is well-draining, enriched with compost, biochar, rice hulls, and oyster flour, contributing to soil health and carbon sequestration. Asia's rice-fish coculture practice is thought to be over 2000 years old. It integrates the production of rice and fish in the same field. It is a symbiotic arrangement in which fish and rice plants benefit.

Organic farming gained momentum in the 20th Century, seeing the negative impacts of industrial farming. In Western countries, Sir Albert Howard, J. I. Rodale, and Masanobu led this movement. In the Indian subcontinent, this was a part of cultural practice. However, in the last few decades, organic farming has increased due to consumer awareness, environmental factors, and health concerns. Realizing the benefits of organic farming, many governments and private organizations have started to support, subsidize, and certify organic products. Many countries have started to integrate organic farming with climate change mitigation policies. This has increased the market for farmers practicing organic farming.

Core Practices of Organic Farming:

- Health: The Basic principle of organic farming is to maintain soil health, plant health, animal health, and human health.
- Ecology: Organic farming practices are in alignment with natural systems rather than against them.
- Care: Organic farming supports sustainability for future generations.
- Fairness: It considers the ethical use of resources to ensure social justice for farmers, workers, and consumers.
- Pest Management: It focuses on biological pest management rather than chemical pesticides. Biological pest control may be achieved with ladybugs and aphids or companion planting, like marigolds, to repel nematodes.

- Certification: Many countries have framed policies and standards for organic farming. For example, USDA Organic or EU Eco-Regulation

Benefits of Organic Farming:

- Soil health: Organic farming enhances soil health through compost, organic matter, and aeration, leading to better water retention, reduced erosion, and nutrient retention.
- Biodiversity: It promotes biodiversity in flora and fauna. It maintains the ecological cycle. Since it does not use synthetic fertilizers or chemical pesticides, it neither degrades the soil quality nor kills the soil organisms that are helpful for plant health. Thus, it also indirectly reduces pest outbreaks.
- Human Health: Perhaps this is the most significant reason people have started to look towards organic farming, as it eliminates the risk of cancer and other life-threatening diseases arising due to the use of synthetic fertilizers and poisonous chemical pesticides.
- Economic Prosperity: Due to the awareness of health and the benefits of organic farming, demand for organic produce has increased worldwide, leading to better opportunities for farmers practicing organic farming.
- Climate Resilience: Due to carbon sequestering in the soil, organic farming helps to reduce global warming and combat climate change.

Challenges of Organic Farming:

- Lower Initial Yields: Compared to conventional farming, the organic farming method may take a few years to reach the optimal yield. This is due to the degradation of soil due to the excessive use of synthetic fertilizers and poisonous pesticides, resulting in eroded, infertile soil.
- Higher labour: Since organic farming does not involve the use of any pesticides, there will be increased manual labour for weeding and pest management.

- Certification Costs: To ensure quality, organic farming needs certification. This may be costly and time-consuming.
- Market Infrastructure: Organic products are generally costlier; they need a separate marketplace to ascertain dedicated customers. This mandates better access and infrastructure for the farmers and consumers of organic products.

Despite these challenges and limitations, various governments and NGOs are trying to promote organic farming to enhance sustainable farming through subsidies, market linkage to farmers, and technical support.

Thinking Corner: The Invisible Farmers

In developing countries, women share 60-80% of farm labour, but own less than 15% of land. Can farming be truly sustainable without bridging the gender gap for equality?

9.3 Permaculture

It is based on the principle of land management and settlement design that emphasizes ecological harmony and sustainability through self-sufficiency. It adopts the natural arrangements of a flourishing natural ecosystem. It mimics the natural ecosystem to achieve the maximum yield. Forest gardening is an example of permaculture.

The basic principles of Permaculture include learning from natural ecosystems to create effective and efficient designs, maximum utilization of natural resources, and minimizing waste. Permaculture techniques include agroforestry, companion planting, contour farming, Hügelkultur (A system of creating raised beds to retain moisture and nutrients using decaying wood), polyculture farming (A system of multiple crops), and biochar implementation (A system of improving soil fertility and carbon sequestration). Sometimes zoning is also preferred, where landscapes are zoned based on their use, for example, zone 1 for herbs and zone 3 for forests.

9.4 Case Study

The success of Organic Farming in India:

India had been practicing organic farming, but to fulfill the growing needs of an ever-increasing population, it too had adopted fertilizers and pesticides. However, many farmers used to follow organic farming in an organized way. The Government of Sikkim set an organized example. To promote organic farming, the Government of Sikkim banned the use of synthetic fertilizers and pesticides. Sikkim became the first state to take this bold step in India. Seeing the results, many other states may soon follow this small state located in the foothills of Mount Himalaya. Sikkim demonstrated the potential and suitability of organic farming on a large scale. The Sikkim government encouraged the farmers to practice organic farming, which has resulted in higher biodiversity, better soil fertility, and far better income for farmers practicing organic farming. Briefly, organic farming is a win-win situation for all stakeholders, including humans, microbes, plants, soil, and the environment.

The Rodale Institute (USA):

They demonstrated that organic systems could match the conventional yields after a five-year transition period and sequester 500 kg of carbon per acre per annum. (source: *Rodale Institute, Farming Systems Trial 2011)*

Loess Plateau Restoration (China):

This project helped transform more than 35,000 square kilometres of eroded land into productive fields, changing the lives of 2.5 million poor people.

Challenges: There is always a debate regarding the yield capacity of organic farming. Farmers fear that by adopting organic farming, their yield will reduce to a meagre 20 to 30 %. However, this is not completely correct. As mentioned earlier, in the initial period, the land will take time to reclaim itself and renew itself, so yield may be reduced, but the selling price will be much higher, so farmers will benefit in the long term. Human health will benefit. Farmers also worry about the intensive labour and the additional expenditure required in accepting organic farming.

Training of farmers will be required to motivate them for this crucial transition. NGOs and governments will have to take proper measures to boost organic farming and permaculture.

However, certain emerging innovations may bridge organic farming's limitations. They may be biochar and mycorrhizal networks to increase the soil carbon storage, using drones to monitor pest hotspots, and the inclusion of the latest technology, like IoT sensors, to detect the maturity of compost.

9.5 Smart Farming

Industry 4.0, the Fourth Industrial Revolution, has contributed many innovative machines to help farmers in many ways. Now it is time to become smart by accepting the concept of Smart Farming. It involves the use of IoT (Internet of Things), AI (Artificial Intelligence), and data analytics to analyse and act in the agriculture field from forecasting the weather, and the time to sow seeds, to assessing the need for irrigating, and the right time of harvesting. We may say it is a revolution from tractors to algorithms. It helps farmers make the right decision based on data to improve crop yield. It includes a network of interconnected sensors to collect real-time data and devices to act in response to the processed signal. Key components may be soil sensors, Climate sensors, drones, satellites, predictive analysis, and machine learning.

9.5.1 Applications of Smart Farming

- Soil and weather monitoring: Sensors are used to measure the moisture, nutrients, level of nitrogen, pH, and temperature of the soil to optimize the irrigation and frequency of fertilizers.
- Smart Pest Control: AI-based systems can detect the growth of diseases in crops at an early stage based on drones, satellite images, smell detection, and image recognition, and provide recommendations for the right treatment.
- Automated Weed Control: Robots interconnected with the system can identify and remove weeds without chemical herbicides.
- Livestock monitoring: With the help of a wearable collar belt, animals' health and location can be traced. Disease prediction and curing are possible at an earlier stage.

- Smart Auto-irrigation System: Based on the temperature and moisture data provided through the soil sensors, an auto-irrigation system may be employed, which automatically starts the water pump or drip lines to water the crops.
- Precision Spraying: It may be applied to spots suffering from certain diseases. This localized treatment helps to recover only affected plants and thus has no side effects on other healthy plants. It also reduces the chances of disease spreading.
- Precision Agriculture: With variable-rate technology, seed waste can be reduced by 30% by adjusting the sowing depending on the soil map.
- Multispectral Imaging: This is a special technique to detect crop stress even before the visibility of disease symptoms. Thus, early disease control is achieved.
- Automated greenhouses: AI-based greenhouses require an ultra-low amount of water and zero pesticides. They are also equipped with control systems to monitor and optimize perfect humidity, temperature, and CO_2 levels.
- Logistics: An AI-driven system may be employed to enhance the supply chain to reduce the waste of perishable food items like vegetables. It can also monitor the cold storage and related infrastructure, including food tracking.
- Predictive Analysis: Forecasting of weather, need for irrigation, Selection of crop based on weather analysis, and optimal harvesting times are possible to have maximum crop yield.

9.5.2 Challenges

- Energy Demand: Sensors, data centres, and IoT networks require enormous amounts of electricity, so it is essential to link them with renewable energy sources to reduce the carbon footprint.
- Digital Divide: Even today, tractors are not accessible to most of the small farmers across the globe, particularly in low-income countries. So, it is very impractical to think of implementing IoT-based smart systems for these

marginalized farmers. Proper infrastructure will be required to be installed, which will require a huge capital investment.

- Data Sovereignty: Often, these IoT-based sensors, bulk data, predictive analysis, etc, are managed by companies or corporations. There will always be a conflict between data privacy and sovereignty between farmers and companies. Clear policies need to be framed.

9.5.3 Case Study

Many countries are doing well using these latest innovations. This is a win-win situation for farmers, consumers, and governments.

- India's Agri Stack is one such program to create a digital ID system for more than 50 million farmers with AI-backed advice through a commonly used messenger system like WhatsApp.
- Kenya has initiated the iCOW project, which sends SMSs related to livestock and crop management.
- FarmBeats from Microsoft is a low-cost IoT kit for offline data transmission. It aims to empower farmers to make better decisions, reduce waste, and increase crop yields with data-driven insights.

9.5.4 The Road Ahead for Smart Farming

More than 85% of farms are owned by small farmers. They do not have access to the latest technology, digital infrastructure, or powerful tools like tractors, threshers, cold storage, and logistics. Market access is also a big issue in many African and Latin American countries. On the other hand, many corporations and companies are already present in the market, and they always have the priority of profit. So, it becomes particularly important to safeguard the interests of these small farmers. There should be a clear policy to strengthen these small farmers. Open-source AI should be promoted to allow farmers easy and free access. Edge Computing is necessary to process data on the farm rather than in the cloud to reduce electricity usage and increase reliability. Algorithms should not be designed just for efficiency but to restore the ecosystem.

9.6 Vertical Farming and Hydroponics

It is estimated that by 2050, more than 68% of the human population will be dwelling in urban areas [44]. People are leaving villages in search of a better life, employment opportunities, education, and status. Our traditional agricultural system will not be able to fulfill the need for food for this vast population. Shrinking farms, increasing arid lands, and increasing industrialization on farmlands are decreasing food production and food security. Forests and wetlands are there, but they should not be encroached on as they are the lungs of our ecosystem and are necessary for our survival and ecological sustainability. New innovative approaches should be used to cope with this challenge. Vertical farming and hydroponics may help to alleviate this problem.

Vertical farming is a method of cultivating and growing plants in stacks arranged in vertical layers. Often, they are in controlled environments and use LED lighting. It is quite easy to control the lighting spectra of LED's red/blue light ratio to accelerate the growth. It is used in urban areas to meet the local demand. Singapore is a small country with limited land. It has motivated people to engage in vertical farming professionally. A Sky Greens company in Singapore has developed a multi-layered vertical farming technique to minimize space while providing fresh vegetables in the vicinity. One will be amazed to read that this company supplies more than 10% of the country's vegetables using this technology.

Advantages of Vertical farming:

- Vertical farming uses much less space in comparison to traditional farming.
- Vertical farming requires 5% of water compared to traditional farming.
- External weather conditions do not affect the crops in vertical farming, as they are in completely controlled conditions.
- Since they are in a controlled environment, pests and insects from around will not be able to penetrate. Hence, no need for chemical pesticides.

- Since they are grown in urban areas or near consumption points, transportation and other logistics costs are reduced. Thus, vertical farming helps to reduce carbon emissions.
- Since they are immune to seasons, year-round production is possible.

Limitations of Vertical farming:

- Vertical farming is an expensive option as it requires LED lighting and controlled environments.
- Vertical farming requires high initial costs. Its starting expenses may exceed Rs 80,000 per square meter; thus, they are not accessible to all.

9.7 Hydroponics

Hydroponics is a special method of growing plants without soil. Plants are grown in water instead of soil. This water is rich in nutrients and minerals. This method often offers a higher yield than traditional farming. They are confined within a small space, so this method is better suited for space-constrained areas. Much water is conserved in this method. Vegetables, fruits, herbs, and flowers can be grown in hydroponics. Hydroponics is a form of sustainable agriculture, when powered by renewables, as it conserves water, has a limited use of land, and does not use pesticides and herbicides. It is used both for small farmers and commercial enterprises. Readers should note that hydroponics is different from aquaponics. Aquaponics involves fish raising, while hydroponics is soilless plant cultivation. So, they should not confuse them with the aqua and hydro words.

> ***Thinking Corner: Feeding 10 Billion!***
>
> *By 2050, without expanding farmland, the world will have to feed 10 billion people. What do you think will be possible to achieve when farmland is shrinking for industries, businesses, and houses? Can smart farming help in this case?*

9.8 Animals: A source of GHG?

Animals other than humans also contribute a lot to global greenhouse gas emissions. The main constituents are methane and nitrous oxide. According to FAO, the livestock sector is responsible for more than 14% of global greenhouse gas emissions [45]. This may be from following:

- Fermentation in ruminants like cows, sheep, and goats (mainly methane).
- Manure decomposition: both methane and nitrous oxide.
- Land use changes: Deforestation, forest burning for agriculture.
- Feed production and transport: Indirect GHG emissions.

Creative Strategies to reduce animal based GHG emissions:

Since a sizeable portion of GHG emissions is contributed by animals, new out-of-the-box creative ideas need to be implemented.

1. A small inclusion of Seaweed (Asparagus Taxiformis) in cattle diet can reduce methane emissions by 80%! This may also restore marine life through the planting of seaweeds.
2. Synthetic biology (e.g., CRISPR-modified gut microbes) may help in genetically modifying microbes of ruminants to produce less methane, by turning emissions into inert gases.
3. Biogas can be generated using bran to capture methane from manure.
4. Selecting climate-resilient low-emission livestock that naturally emit less methane.
5. Lab-grown meat and cultured dairy products can significantly reduce animal-related GHGs.
6. Rotational grazing and reduction can improve soil carbon storage.
7. Reduced use of synthetic fertilizers reduces emissions of nitrous oxide.
8. Methane-absorbing nanomaterials can also be helpful.
9. Manure can be converted into biochar (a carbon-rich soil additive) to sequester carbon for the long term in soils.

FUN FACT: ***Do you know, Denmark has decided to impose a tax on livestock emissions from 2030?***

> ***Thinking Corner**: You know very well that Denmark is the leading producer of milk and milk products. But these cattle don't just give milk, they also release cow dung! 1,15,80,000 pigs, 11,40,000 Bovine cattle, 5,47,000 Dairy cows, and 1,33,000 sheep produce 3-4 crore tonnes of dung. So, a mountain of dung is produced every year.*
>
> *Suppose you are the prime minister of Denmark. What actions will you take to solve this issue?*

9.9 Desert Forestation: Can this be helpful?

Afforestation and desert forestation can help absorb carbon at the fastest rate. So, the best way is to increase the green cover on the planet. Urban forests may be increased in the cities, roadsides, streets, rooftops, open spaces available in and around our homes, offices, schools, malls, and other buildings. One can find plenty of space in streets, verandas, roadsides, farms, and public places. Unlike rural and urban areas, deserts have a scarcity of water, high temperatures, and extreme weather, but plenty of land! Extreme climatic conditions are unfavourable for the survival of plants. The following are a few strategies to counter it:

1. Controlled microclimate shields plants from heat, sandstorms, and wind. It also helps in maintaining humidity and a stable temperature.
2. Drip irrigation can reduce water requirements.
3. A greenhouse can reduce water loss.
4. Saline water from the sea can be used after desalination or can even be sourced from recycling facilities.
5. Maximum protection is needed in the initial 3 to 5 years. Once plants mature into trees, they can survive on their own. So, special precautions are required during the first 5 years.
6. Genetically modified, heat-resistant, water-efficient species should be planted to maximise the green cover in deserts. Once greenery reduces the ambient temperature and sandstorms, we may plant a normal plantation.
7. Desert native drought-tolerant plants like Acacia, Jatropha, Moringa, and Prosopis may be planted in large areas.

8. Solar-powered sensors and drip irrigation may be used for automation. Small modular reactors (SMR) may be deployed for desalination plants to meet water requirements.

Real World Examples:

- Sundrop Farms is an excellent example of forestation using a greenhouse in the desert. It is powered by solar and watered by the desalination of seawater.
- Another example is from Qatar and Jordan. Sahara Forest Project is powered by CSP solar. It works for the afforestation in the arid zones of the Sahara Desert. *(Sahara Forest Project Report, 2022)*

> ***Thinking Corner: Dessert to Forest***
> *Is it possible to convert all deserts of world into forests using controlled greenhouse?*
> *What may be the positive and negative consequences?*

9.10 Conclusion

The moral of sustainable agriculture can be summarized in a single sentence: *"We do not inherit the land from our ancestors but borrow from our children" -A Native American Proverb..* Sustainable agriculture is the judicious use of land, water, food grains, and fertilizers. It advocates for organic farming, vertical farming, permaculture, and hydroponics for better utilization of present resources to maintain them even for future generations, while caring for our health. It also emphasizes the integration of modern innovations of Artificial Intelligence, IoT sensors, satellite imagery, and weather forecasting to increase the yield, lower expenses, and improve pest management.

-:EXERCISE:-

Section A: Short Answer Questions

1. Define Sustainable Agriculture.
2. List two goals and two benefits of sustainable farming.

3. List two traditional practices aligning with sustainable agriculture.
4. State how crop rotation helps improve soil health.
5. Define Organic Farming.
6. List the negative impacts of chemical fertilizers.
7. Define Permaculture.
8. State the role of vermin composting in sustainable agriculture.
9. List two challenges in adopting sustainable agriculture in India.
10. List the benefits of Organic farming.

Section B: Long Answer Questions

1. Discuss the role of sustainable agriculture for soil and human health.
2. Explain the major practices of sustainable farming.
3. Compare conventional farming with organic farming.
4. Describe the role of mixed farming in maintaining biodiversity and soil health.
5. List the advantages of organic farming.
6. Analyse the impact of climate change on agriculture. Mention the mitigation strategies for the same.
7. Design a model of sustainable farming for water-deficient areas, incorporating crop planning and renewable energy.
8. Discuss the role of policy and frameworks in promoting sustainable agriculture.
9. List the benefits of vertical farming.
10. Discuss the role of organic farming in human health.

Section C: Multiple Choice Q and A

1. The global population was around ....by the end of 2022.
   a) 7 billion b) 8 billion c) 9 billion d) 10 billion
2. The human population took… to grow from 1 billion to 8 billion.
   a) 118 years b) 218 years c) 518 years d) 918 years

3. The current annual addition to the global population is:
   a) 50 million b) 100 million c) 80 million d) 200 million
4. By the end of 20510, the human population may reach:
   a) 8.7 billion b) 9.7 billion c) 10.7 billion d) 11.7 billion
5. The primary issue regarding food production in the future is:
   a) Excess production
   b) Scarcity of fertile land and water
   c) Lack of demand
   d) Overuse of organic farming
6. The main goal of sustainable agriculture is:
   a) Maximize profits
   b) Meet present needs without compromising future generations
   c) Focus on monoculture
   d) Use only synthetic fertilizers
7. …….. is NOT a principle of organic farming:
   a) Use of synthetic pesticides
   b) Soil health maintenance
   c) Natural pest control
   d) Biodiversity promotion
8. The "Terra Preta" is:
   a) A type of chemical fertilizer
   b) Fertile soil enriched with biochar and compost
   c) A hydroponic farming technique
   d) A genetically modified crop
9. The ………… state of India has fully adopted organic farming.
   a) Kerala b) Gujarat c) Sikkim d) Punjab
10. The organic farming had a major challenge of:
    a) Lower labor costs b) Higher initial yields
    c) Certification expenses d) Excessive water usage

11. A permaculture emphasizes:
    a) Monoculture farming b) Ecological harmony and self-sufficiency
    c) Industrial farming d) Heavy pesticide use
12. The… technique uses raised beds of decaying wood.
    a) Agroforestry b) Hügelkultur

c) Polyculture d) Hydroponics

13. The ……. A project in China transformed eroded land into productive fields.
    a) Organic Mission b) Loess Plateau Restoration
    c) Agri Stack d) Farm Beats
14. About organic farming, Rodale Institute (USA) showed that:
    a) It always has lower yields
    b) It can match conventional yields after a transition period
    c) It increases pesticide use
    d) It is only suitable for small farms
15. Smart farming primarily uses:
    a) Traditional tools b) IoT and AI
    c) Chemical fertilizers d) Only manual labor
16. …….. is NOT an application of smart farming.
    a) Soil monitoring b) Livestock tracking
    c) Manual weeding d) Predictive weather analysis
17. A major challenge of smart farming is:
    a) Low energy demand b) Digital divide and prohibitive costs
    c) Lack of government support d) Excessive water use
18. The ………… Indian initiative provides AI-backed advice to farmers via WhatsApp.
    a) FarmGPT b) Farm Beats
    c) Sky Greens d) iCOW
19. A key advantage of vertical farming is:
    a) High water consumption
    b) Immunity to external weather
    c) Only suitable for rural areas
    d) Requires a large land area
20. ………. country is a leader in vertical farming.
    a) India b) Singapore c) Brazil d) USA
21. In hydroponics, plants are grown without:
    a) Water b) Soil c) Nutrients d) Light
22. ……… is a limitation of vertical farming.
    a) Low initial cost b) High energy consumption
    c) High pesticide use d) Limited crop variety

23. ……. The percentage of global GHG emissions comes from livestock.
    a) 5% b) 14% c) 25% d) 50%
24. ……. Gas is primarily emitted by ruminants like cows.
    a) Nitrous oxide b) Methane c) CO_2 d) Oxygen
25. The methane emissions from cattle can be reduced using:
    a) Feeding them seaweed,
    b) Increasing synthetic fertilizers
    c) Using more pesticides,
    d) Reducing water intake
26. A key strategy for desert afforestation is:
    a) Using excessive water,
    b) Drip irrigation, and Microclimate control
    c) Avoiding native plants,
    d) Heavy pesticide use
27. …….. project uses solar power and seawater desalination for farming in deserts.
    a) Agri Stack b) Sahara Forest Project
    c) iCOW d) Farm Beats
28. Sustainable agriculture teaches us:
    a) Maximize short-term profits
    b) Borrow resources from future generations responsibly
    c) Focus only on high-yield crops
    d) Use only chemical fertilizers
29. **T/F:** The global population grew from 1 billion to 8 billion in 218 years.
30. **T/F:** Synthetic fertilizers are used in organic farming.
31. **T/F:** Sikkim became the first Indian state to adopt organic farming.
32. **T/F:** The traditional farming uses less water than vertical farming.
33. **T/F:** In hydroponics, plants are grown in soil-rich environments.
34. **T/F:** Livestock contributes over 14% of GHG emissions.
35. **T/F:** Smart farming uses IoT and AI but does not involve predictive analysis.
36. **T/F:** In China, the "Loess Plateau Restoration" project transformed the eroded land.

37. **T/F:** No water management strategies are required in desert afforestation.
38. **T/F:** To maximize yield, natural ecosystems are mimicked in permaculture.
39. Overpopulation leads to .... of resources fast.
40. _---- farming avoids synthetic fertilizers.
41. Amazon's fertile __________ soil was created over 2,000 years ago using biochar.
42. In Singapore, __________ is a vertical farming company that supplies 10% of the country's vegetables.
43. The soil moisture and nutrients in smart farming are monitored using........
44. Methane emissions can be reduced significantly by adding... in cattle feed..
45. The __________ project of Kenya for livestock management sends SMS alerts.
46. The plants are grown in nutrient-rich water without soil in ....
47. The .... involves maximizing natural resource use.
48. .... farming stacks crops vertically in controlled environments.
49. The solar power and seawater desalination for farming are used in .... Project.
50. Sustainable agriculture aims to borrow resources from __________ generations responsibly.

Answer Keys:

1.B, 2.B, 3.C, 4.B, 5.B, 6.B, 7.A, 8.B, 9.C, 10.C,
11.B, 12.B, 13.B, 14.B, 15.B, 16.C, 17.B, 18.A, 19.B, 20.B,
21. B, 22. B, 23. B*, 24. B, 25. A, 26. B, 27. B, 28. B, 29. T, 30. F, 31.T, 32. F, 33. F, 34. T, 35. F,36. T, 37. F, 38. T, 39. depletion, 40. Organic, 41. Terra Preta, 42. Sky Greens, 43. sensors, 44. Asparaguses taxiformis (seaweed), 45. iCOW,
46. Hydroponics, 47. Permaculture 48. Vertical, 49. Sahara Forest, 50. future

**FAO 2023 estimates range: 11-14%*

चरैवेति, चरैवेति।

Aitareya Upanishad, a part of the Rigveda (7.15.14)

Move forward, keep moving!

Chapter 10

SUSTAINABLE TRANSPORTATION

Learning Outcomes:

After reading this chapter, the learner will be able to:

1. Understand the role of traditional transportation in degrading the climate.
2. Explore the role of EVs and hydrogen vehicles in mitigating climate change.
3. Discuss the smart public transportation system.
4. Explore the future trends in transportation systems.

10.1 Introduction

Transportation refers to the movement of man and material from one location to another. In ancient times, people used animals for transportation. After the Industrial Revolution, trains, tractors, buses, cars, ships, and planes were used for transportation. In most cases, the manufacturing location of goods used to be far away from the point of consumption, so transportation is required. More than 30% of our energy is consumed in the transportation sector [3]. Road transport is the largest consumer of energy, followed by air transport. Water transport consumes a minimum amount of energy. Shell has estimated a 60% rise in demand in the global transportation sector [46]. The following graph indicates the rise in energy demand of the transportation sector in the last fifty years.

As the world faces pollution, global warming, and climate change, transportation has become a significant sector. A more efficient, economically viable, and sustainable solution is required for the transportation sector. Sustainable transportation refers to the energy-efficient movement of humans, animals, goods, and materials with minimum impact on the environment. It includes many options like electric vehicles, hydrogen-powered buses and trains, energy-efficient smart public transit systems like metro, and hyperloops. Soon, it may include drones and air taxis. Transitioning to these clean transportation methods can significantly reduce our carbon footprint.

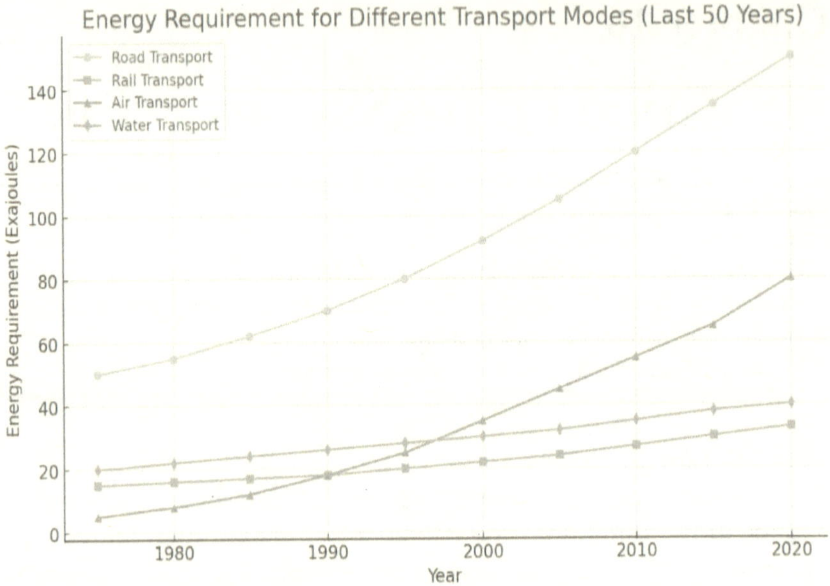

Figure 10.1 Energy Requirement for Road, Rail, Air, and Water Transportation Systems

This can further reduce air pollution and improve public health. Since most of the vehicles used in the transportation sector use petrol or diesel as fuel, this sector is one of the largest generators of greenhouse gases. However, many countries are now slowly switching towards electric vehicles to reduce their carbon footprint.

The sustainable transportation sector not only contributes to the environment but also helps in the generation of employment opportunities in the manufacturing and maintenance industries. It also promotes public transportation, leading to reduced traffic congestion. Additionally, sustainable smart public transportation can help in bringing social equality through affordable, reliable mobility for all, despite different socio-economic backgrounds. It can also contribute to a healthy environment by reducing air and noise pollution. Even when sustainable transportation has many advantages as mentioned above, it has a few challenges too! At present, many countries are promoting electric vehicles as an alternative to traditional vehicles, but we lack the required charging infrastructure for the same. Charging time discourages many customers from purchasing the EVs. Much innovation is required in the field of batteries to increase the range. On the other hand, hydrogen-powered transport promises a new dawn in the non-polluting, sustainable transportation sector, but it needs heavy investment and research to make it economically viable. Self-driving taxi, air taxi, and drone-based delivery systems are a new field of transportation, but a policy framework is missing to regulate them.

The sustainable transportation sector can blossom if governments, industries, and individuals work collectively. For example, the government can frame positive policies, provide subsidies to electric vehicles, charging stations, and research for EVs and hydrogen-powered vehicles. Industries may plan research on innovative ideas to increase the life and range of batteries, to reduce the weight of batteries, and search for cost-effective mass manufacturing. Individuals may adopt electric vehicles in place of traditional fossil fuel-based vehicles and maximise the use of the public transportation system.

In the following sections, we shall discuss in detail the role of electric vehicles and hydrogen-powered vehicles in sustainable transportation. The smart public transportation system has also been explained to appreciate its role in addressing environmental issues.

10.2 Electric Vehicles

Electric Vehicles are rapidly replacing the traditional internal combustion (IC) engines. EVs do not produce greenhouse gases like traditional IC engines. Thus, they are eco-friendly. They run on electricity, which may be procured from renewable energy sources like wind and solar. EVs are adopted worldwide to reduce the impacts of climate change and global warming, and to conserve the environment. Almost all developed and most developing nations are supporting EVs through positive policies and subsidies [27]. In this section, we shall discuss the rise of EVs, advantages and limitations of EVs, charging infrastructure, novel technologies in batteries, government policies, and associated challenges to accept EVs as common, affordable vehicles.

- Rise, Fall, and Rise of Electric Vehicles: Today's generation may be surprised to know that EVs were discovered before petrol and diesel engines. The history of electric vehicles is more than a century old. In 1828, which means almost two hundred years back, Hungarian engineer Ányos Jedlik first built a primitive model of EV comprising a small electric motor to move a small car model. In 1881, Gustave Trouve operated a tricycle for the first time with the help of a rechargeable lead-acid battery. In 1890, William Morrison built a car with the capacity of six passengers, having a maximum speed of 22 km per hour. In 1899, Camille Jenatzy built a car that could reach a speed of more than 100 km/h. It was the fastest speed ever witnessed by humans on roads. These electric vehicles are gaining place in people's minds due to their modest maintenance, no noise and zero emissions. [23].

  After the invention of petrol and diesel engines, which used the internal combustion technology, they were very convenient to operate. Petroleum products were much cheaper. Electric vehicles were having problems with long charging times and had limited range. On the other hand, technology and the market of IC engines made it much affordable. So, people started to prefer the IC engine-based petrol and diesel vehicles for transportation and movement. Electric vehicles were almost forgotten for half a century, from 1940 to 1990.

  Since 1990, pollution has started to become one of the major problems. People and the government started to realise the ill effects of global warming and climate change. This forced the technological sector to rethink electric vehicles, which do not produce greenhouse gas and are environmentally friendly. In 1996, General Motors manufactured the *EV1* model car; later in 1997, Toyota introduced the *Prius*, a hybrid car. These initial models boosted the interest, need, and awareness for electric vehicles.

The real EV revolution started in the 21st century. In 2008, Tesla introduced the *Roadster* car, which had a higher range and higher performance. In 2010, Nissan introduced the *Leaf*, which was the world's first mass-market car. Since 2015, global leaders in automobiles like General Motors, Tesla, Ford, Volkswagen, Tata, Suzuki, and many others have geared up for innovations in electric vehicles and associated areas like charging infrastructure, batteries, and power electronics.

> ***Thinking Corner:*** *Do you know, a diesel bus carrying 40 passengers is GREENER than 10 electric cars carrying the same people?*
>
> *Is shifting to public transport more impactful than shifting to green fuel?*
>
> *What do you and your parents prefer, a personal car or public transport?*

- **Working of Electric Vehicles:**

  The basic principle involved in almost all electric vehicles is the conversion of stored electrical energy into mechanical energy to propel the vehicle. Essential components of EV are the Electric Motor, battery pack, Power Electronics Converter, and Charging System. Unlike IC engines, multiple gearboxes are not used in the case of EVs. Regenerative braking is used to generate electricity during braking and travelling through a descending gradient. The onboard battery pack is charged through a rectifier, which converts AC into DC. Inverters and power electronics converters are used to provide a controlled power supply to an electric motor, which converts electrical energy into rotational mechanical energy, necessary to propel the cart. The following two diagrams represent the energy flow during forward movement and breaking.

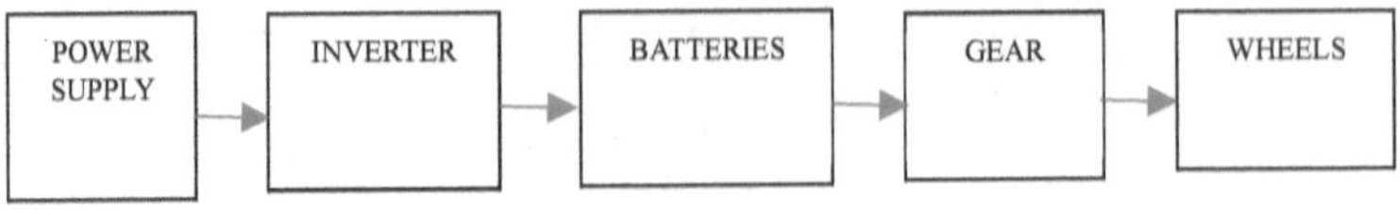

Figure 10.2 Schematic Diagram of Electric Vehicles

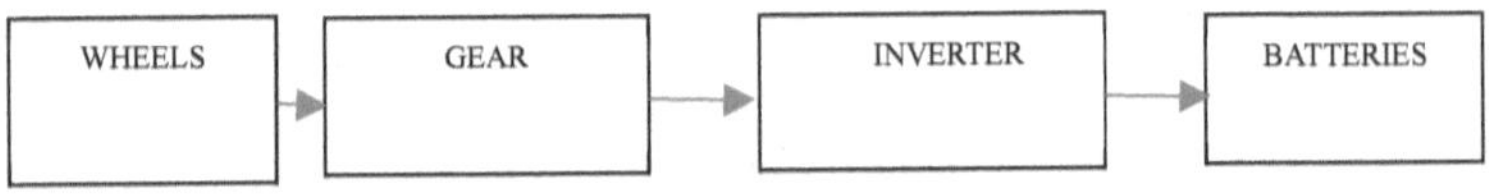

Figure 10.3 Schematic Diagram of Regenerative Braking in Electric Vehicles

- **Benefits of Electric Vehicles**
  1. The biggest benefit of electric vehicles is the pollution reduction. EVs are environmentally friendly, leading to a cleaner environment. If they are charged with renewable energy sources, then their carbon footprint is reduced to almost zero.
  2. Electric vehicles are more efficient than traditional internal combustion engines.
  3. Electric vehicles are less complex than other traditional locomotive options. They have very few components, like a battery, a charger, and a motor.
  4. Electric vehicles have lower operating costs.
  5. Since electric vehicles have fewer components, their maintenance is also reduced.
  6. Electric vehicles provide instant torque, thus smoother acceleration and deceleration, leading to a smoother ride.
  7. Electric vehicles produce little noise.

8. Electric vehicles also have the facility of regenerative braking. Thus, it can produce power while reducing the speed or while moving down slopes.

- **Battery: A bottleneck for EVs:**
Batteries have always been a major issue in the mass adoption of electric vehicles. People always expect a higher range, lower charging time, and higher charge-to-weight ratio. Initially, nickel-cadmium batteries were used, but they are very heavy. Now, the most common type of battery used in the world is Lithium-ion batteries. Lithium-ion batteries have a lower weight, high energy density, higher efficiency, and longer durability, but they are costlier. Lithium stock on Earth is limited, and its mining process is also challenging. However, advancements in battery technology and chemistry have reduced costs and improved the range and charging times. Solid-state batteries are being explored to get higher energy density at lower costs. Lithium-sulphur batteries are also being explored for enhanced range and cost reduction. Recycling and repurposing of batteries is very necessary to reduce the pollution caused by battery chemicals. These chemicals may contaminate water and are harmful to our health.

- **Charging Infrastructure:**
Apart from batteries, charging infrastructure is also a significant hurdle in adopting electric vehicles as a common locomotive for transportation. Mainly, there are three types of chargers:
    1. Type 1: It is used for small electric vehicles and has a household outlet of 220 volts. It provides slow charging, but vehicles that are kept on charge may get fully charged after overnight charging.
    2. Type 2: It also has a 220-volt charging structure and is generally available at workplaces, homes, and charging stations. This type of charging station has lower charging times than type 1.
    3. Type 3: In this type, the DC fast charging method is used. It used direct current to charge EVs above 80%

within an hour. These are suitable for long-distance travel.

- **Incentives:** Since electric vehicles are considered to be environmentally friendly, many countries across the world are promoting EVs as a common, affordable vehicle to be adopted by common people. To make EVs more accessible to consumers, governments give tax credits, rebates, and loans. Higher standards to limit air pollution in IC engines mandate that the auto companies shift to electric vehicles. To ensure easy availability of a charging network, public and private charging stations are encouraged.
- **Challenges:** Electric vehicles have many benefits, but this sector is also facing several challenges to penetrate the market. As said earlier, the major bottleneck in the adaptation of EVs is the battery. The main material of the battery is lithium-ion, which has mining concerns and environmental issues. Expansion of public charging stations/points requires a huge investment and proper management. To take the best advantage of EVs, it is also necessary to expand the renewable energy generation. Existing grid infrastructure is also required to be enhanced to meet the demand for electrical energy from the transportation sector. The government will have to aggressively take steps for the mass adoption of electric vehicles. The future of EVs is promising, with continuous improvements in battery technology, government policies promoting clean transportation, and increasing consumer demand driving the transition away from fossil fuels. As global efforts to combat climate change intensify, EVs are expected to play a pivotal role in sustainable mobility.

10.3 Hydrogen-Powered Transport Systems

A hydrogen-powered transport system is one of the best-suited alternatives to traditional fossil fuel energy sources. It offers clean energy, which is non-depleting in nature. It is also sustainable in nature, as the main resource required is water. It produces electricity through an electrochemical reaction between hydrogen and oxygen. Water is produced as the output. No poisonous smoke or any other polluting chemicals are generated during this process. The following section describes the technology, merits, and limitations of this technology in detail.

- **Fuel Cell Technology**: The Fuel cell is the heart of hydrogen power. It converts chemical energy into electrical energy. Hydrogen and oxygen react to form water vapour while liberating energy. The proton-exchange membrane fuel cell is the most common type of fuel cell technology used in transportation applications due to its higher efficiency and ability to function even at low temperatures. The main components of this technology are:
    1. Hydrogen storage tank: It is used to store the compressed hydrogen gas at pressures ranging from 350 to 700 bars.
    2. A stack of Fuel cells: They are used to produce electrical energy to propel the vehicle.
    3. Electric motor: It uses the electrical power generated by fuel cells to produce mechanical energy to drive the vehicle.
    4. Batteries: These are used in parallel to fuel cells to optimise the efficiency for emergency operation and services.
- **Advantages**: Hydrogen-powered transport systems have many advantages in comparison to traditional fossil fuel technology as well as battery-electric vehicles. A few are listed below:
    1. The main and most admired advantage of hydrogen technology is zero emissions. It only produces water vapour in the exhaust. It does not produce any harmful emissions. Thus, they are eco-friendly.
    2. Hydrogen-based transport vehicles have a higher range. Thus, it relinquishes the driver from range anxiety. So, they are more suited for applications involving longer routes.
    3. Hydrogen-based vehicles do not require long refuelling times like battery-operated

vehicles, which need hours to recharge the batteries. This makes them an ideal choice for large commercial fleets.

4. Since hydrogen can be easily compressed in storage tanks and is easy to transport, it offers a potential advantage to store the excess renewable power generated during the low-demand hours and utilise it during the high-demand hours.

- **Infrastructure development:** A huge amount of investment will be required to develop the hydrogen refuelling infrastructure. It requires special equipment and heavy storage devices capable of withstanding high pressure. Hydrogen can be produced with many different methods, like steam methane reforming, electrolysis, and biomass gasification. However, hydrogen produced using renewable energy sources is the most suitable option from a climate change perspective and sustainability. It is also known as Green hydrogen. Special facilities are required to be built for the transportation of compressed or liquefied hydrogen. Many countries, including Germany, Japan, and South Korea, have started to invest in the hydrogen economy, considering it as a fuel for the future. In these countries, hydrogen stations are being built to provide easy options for refuelling. Not just these countries, governments across the world have started to understand the potential and importance of hydrogen as a fuel for the future and are offering many subsidies and incentives for hydrogen production, refuelling stations, and the related infrastructure.
- **Challenges:** Hydrogen is indeed considered a fuel for the future due to its non-polluting exhaust, but many challenges must be overcome to make hydrogen an alternative to traditional fossil fuels. These challenges include high production costs of hydrogen, infrastructure limitations, and lower overall efficiency of hydrogen-driven systems in comparison to battery-operated vehicles. Due to specialised equipment to handle high-pressure hydrogen gas, refuelling stations are also costly to build. To ensure safety, high standards are required to be observed. However, as the

renewable energy sector expands, the cost of producing green hydrogen is reducing to make hydrogen an economically sustainable energy source.

10.4 Smart Public Transportation System

The public transportation system should be strengthened to reduce greenhouse gas emissions. It also reduces congestion, traffic, and noise on the road. The public transportation system is always considered a lifeline of any urban area. The city train system of Mumbai is playing a vital role in the development, communication, and business of Mumbai and the peripheral areas. However, in the last few decades, people have become more inclined towards personal vehicles like cars. The following figure shows the number of cars worldwide in the last 25 years.

This increase in personal vehicles for commuting may be due to flexibility, ease, comfort, less access to public transportation, or status, but it increases traffic and pollution. Seeing the ill effects of global warming, many governments across the world have now started to promote the smart public transportation system.

A smart public transportation system integrates the latest technological advancements, smart technologies, the internet, and IoT to enhance the efficiency, adaptability, and sustainability of the transportation network. It includes real-time data analysis, digital algorithms to optimise the route, traffic, and passenger management. Digital payment systems have streamlined the long, tiring queues of ticketing, reduced wait times, and improved the overall experience of passengers as well as staff. The last decade has also witnessed a rapid transformation in app-based cab booking, which are offering more personalised services without owning cars.

Urban authorities are now electrifying their transit fleets with city buses, electric and hybrid buses, rather than conventional diesel buses. This move has not just reduced their operational costs but also contributed to environmental sustainability.

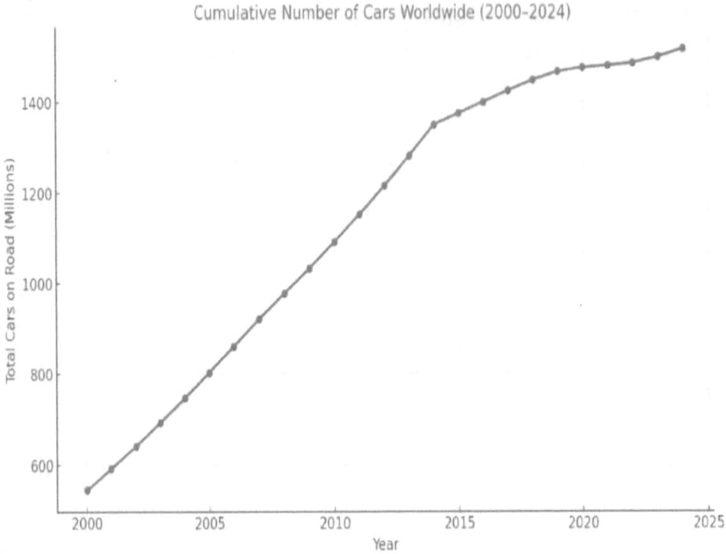

Figure 10.4 Total Number of Cars in the last 25 years (globally)

Bullet Trains and High-Speed Rails: Many countries, including Japan, China, France, and Germany, have invested in high-speed trains. Trains are the most efficient method of transportation. They run on electricity and produce lower emissions than air travel. They provide a fast, reliable, and environmentally friendly way of transportation. Japan's Shinkansen, China's Fuxing and Europe's TGV are a few of the best examples of high-speed rail that have been appreciated by people. They are known for their fast and comfortable services. China's Maglev trains are currently the fastest commercially operational trains, with a speed of up to 431 km/hr.

Hyperloop technology is considered to be a futuristic mode of transport. Near-vacuum tubes are used to help hyperloop pods travel at speeds above 1000 km/hr. This technology consumes a low amount of energy. Virgin Hyperloop and Boeing Company are pioneers in this area. If successful, then it can revolutionise transportation by significantly reducing travel time as well as carbon footprint.

Many companies like Airbus, Boeing, and ZeroAvia are exploring electric and hydrogen-powered aircraft for urban transportation, which can provide taxi services for passengers and cargo services for materials. Electric vertical take-off and landing aircraft (eVTOL) is also a new concept being tested by Joby Aviation, Volo Copter and Archer for urban air taxi services. However, regulations and policies from authorities may be deciding factors for the mass applicability of this sector.

Green shipping and sustainable maritime transport refer to the utilisation of alternative fuels such as hydrogen, ammonia, biofuels, and liquified natural gas (LNG). It may also include ships using battery electric systems and wind assistance for propulsion. Since major cargo across the world is transported through marine ships, even a little increment in the efficiency of ships significantly decreases the greenhouse gas emissions. Maersk and Hyundai Heavy Industries are leading companies exploring zero-emission ship designs.

Activity: Observe the number of people taking trains every day for commuting. Aren't our trains too crowded? Indian trains carry 8 billion people annually yet are underfunded in comparison to the roads and cars. Which should be given the push: mass transit or personal EVs?

-:EXERCISE:-

Section A: Short Answer Questions

1. Define Sustainable Transportation.
2. List three benefits of electric vehicles.
3. Mention the benefits of public transportation.
4. List the benefits of a smart mobility system.
5. State how hydrogen fuel contributes to sustainable transportation.
6. State the significance of cycling and walking in sustainability.
7. List two challenges in adopting electric vehicles in a developing country like India.

8. State two advantages of the Bus Rapid Transportation System (BRTS) in Metro cities.
9. List the main components of an electric vehicle.
10. Arrange road, air, rail, and marine transportation systems in decreasing order of GHG emissions.

Section B: Long Answer Questions

1. Explain the role of sustainable transportation in mitigating climate change.
2. Discuss the social and environmental impact of shifting from fossil fuel-based transport to a clean alternative.
3. Compare electric and hydrogen vehicles from different dimensions affecting environmental sustainability.
4. Discuss the role of ICT in the transportation sector.
5. Explain how government policies can promote sustainable transportation.
6. List various challenges in developing infrastructure for electric vehicles in India.
7. List various activities to reduce the carbon footprint for transportation in your city.
8. Design a sustainable transportation plan for the metro city of Mumbai to get rid of traffic congestion, incorporating EVs, metro trains, and other smart mobility solutions.
9. Evaluate the potential of biofuels as an alternative fuel for sustainable transportation.
10. Explain in detail the working principle of electric vehicles. Mention the advantage of regenerative braking.

Section C: Multiple Choice Q and A

1. The transportation sector consumes ……… percentage of global energy.
   a) 10% b) 20% c) 30% d) 50%
2. The ……… transport mode consumes the least energy.
   a) Road b) Rail c) Water d) Air
3. The estimated rise in global transportation demand by 2050 is:
   a) 30% b) 60% c) 90% d) 10%

4. .... is not a sustainable transport option.
   a) Electric vehicles b) Diesel trucks
   c) Hyperloop d) Hydrogen buses
5. EVs were first built in:
   a) 1800 b) 1828 c) 1928 d) 1978
6. The first mass-market EV was introduced by:
   a) Tesla b) Nissan
   c) General Motors d) Toyota
7. Regenerative braking in EVs helps to:
   a) Reduce weight b) Recharge the battery
   c) Increase speed d) Improve aerodynamics
8. The most common type of battery in modern EVs is:
   a) Nickel-Cadmium b) Lithium-ion
   c) Lead-Acid d) Solid-state
9. The fastest EV charging type is:
   a) Type 1 b) Type 2
   c) Type 3(DC Fast Charging) d) Wireless
10. Governments promote EVs through:
    a) Tax credits b) Banning old vehicles
    c) Banning public transport d) Higher fuel taxes
11. Solid-state batteries aim to improve:
    a) Charging speed b) Energy density
    c) Vehicle weight, d) Cost
12. A major bottleneck in EV adoption is:
    a) High speed b) Battery limitations
    c) Lack of roads d) Noise pollution
13. The emission of hydrogen-powered vehicles is:
    a) CO_2 b) Water vapour
    c) Nitrogen oxides d) Methane
14. ...is the heart of hydrogen transport technology.
    a) Rectifier b) Fuel cell
    c) Inverter d) Lithium battery
15. Hydrogen fuel cells convert:
    a) Chemical energy to electrical energy
    b) Mechanical energy to heat
    c) Wind energy to motion
    d) Solar energy to electricity
16. An advantage of hydrogen vehicles over EVs is:
    a) Faster refuelling b) Smaller size

c) Lower cost d) No oxygen requirement

17. Green hydrogen is produced using:
a) Coal b) Renewable energy
c) Nuclear power d) Natural gas
18. …….. is a major challenge for hydrogen transport.
a) High production cost b) Low energy density
c) Short engine life d) Excessive emissions
19. …….. countries are investing heavily in hydrogen infrastructure.
a) India and Brazil b) Germany and Japan
c) Australia and Canada d) USA and Russia
20. Smart public transport systems use:
a) IoT and real-time data b) Light vehicles
c) Manual ticketing d) Only diesel engines
21. The fastest commercial train operates in:
a) Japan b) China c) Germany d) France
22. Hyperloop technology uses:
a) Near-vacuum tubes b) Solar-powered tracks
c) Magnetic roads d) Underwater tunnels
23. eVTOL stands for:
a) Electric Vehicle Taxi on Land
b) Electric Vertical Take-off and Landing
c) Eco-friendly Vehicle Technology
d) Efficient Vehicle Transport Overload
24. Green shipping uses fuels like:
a) Hydrogen and ammonia b) Coal-based fuels
c) Only diesel d) None of the above
25. The growth of EVs declined in the 20th century due to:
a) Cheaper petroleum b) High EV speeds
c) Lack of people's interest d) Lack of pollution concerns
26. ………. is not a benefit of EVs.
a) Energy efficiency b) Higher noise pollution
c) Reduced maintenance d) Lower emissions
27. ……. Is the main challenge in hydrogen refuelling.
a) High-pressure storage b) Blast
c) Lack of oxygen d) Slow refuelling
28. The primary goal of sustainable transport is to:
a) Increase car sales b) Reduce carbon footprint
c) Stop public transit d) Ban fossil fuels.

29. ……….. energy source powers the most sustainable EVs.
    a) Coal b) Renewable c) Nuclear d) Natural gas
30. …….. company is a pioneer in the hyperloop.
    a) Virgin Hyperloop b) Tesla
    c) Boeing d) Airbus

True/False

31. Transportation consumes over 30% of global energy.
32. Water transport for freight is costlier than air transport.
33. EVs were more popular than petrol cars in the 1920s.
34. Regenerative braking charges EV batteries.
35. Type 3 EV chargers are slower than Type 1.
36. Hydrogen vehicles' refuelling is slower than EVs.
37. Japan has invested heavily in hydrogen infrastructure.
38. IoT and AI are used in smart transit systems.
39. Digital payments reduce public transit efficiency.
40. All countries have adopted hyperloop technology.
41. Shell estimates a ___% rise in transportation demand.
42. The Nissan ___ was the first mass-market EV.
43. ___ braking recharges EV batteries during deceleration.
44. _....... hydrogen is produced using renewables.
45. Hydrogen is stored at ___ bars of pressure.
46. ___ trains are the most energy efficient.
47. Hyperloop pods travel in …….. tubes.
48. The main bottleneck for EVs is ……… technology.
49. Recycling EV batteries reduces ___ pollution.
50. ___ trains reduce travel time and emissions.

Answer Keys:

1.C, 2. C, 3. B, 4. B, 5. B, 6. B, 7. B, 8. B, 9. C, 10. A, 11. B, 12. B, 13. B, 14. B, 15. A, 16. A, 17. B, 18. A, 19. B, 20. A, 21. B*, 22. A, 23. B, 24. A, 25. A, 26. B, 27. A, 28. B, 29. B, 30. A

True/False:
31. T, 32. F, 33. F, 34. T, 35. F, 36. F, 37. T, 38. T, 39. F, 40. F

Fill-in-the-blanks:
41. 60, 42. Leaf, 43. Regenerative, 44. Green, 45. 350 to 700, 46. High-speed rail, 47. Near-vacuum, 48. Battery, 49. Chemical, 50. Bullet

**Japan's L0 Maglev (603 km/h) holds the speed record but is not yet commercial.*

एवं प्रवर्तितं चक्रं नानुवर्तयतीह यः।
अघायुरिन्द्रियारामो मोघं पार्थ स जीवति॥

(Bhagavad Gita 3.16)

"O Parth! One who does not follow the divine cycle of creation, consumption, and renewal and lives merely for sensory enjoyment lives in vain."

Relevance:

The above shloka from the Bhagavad Gita warns us against the overconsumption and unsustainable exploitation of natural resources to align with limited industrial production (read here carbon footprint). It also motivates us to judiciously use the resources to reduce the impact on the environment (a central theme of green manufacturing). It suggests practicing ecological balance through zero waste and the circular economy.

Chapter 11

GREEN MANUFACTURING AND WASTE MANAGEMENT

Learning Outcomes:

After reading this chapter, the learner will be able to:

1. Understand the role of green manufacturing in sustainable Development.
2. Apply the principle of circular economy (that is, 5R: Refuse, Reduce, Reuse, Repurpose, and Recycle).
3. Discuss the energy management system (EnMS) to implement eco-friendly industrial processes.
4. Evaluate the benefits and challenges of an energy audit.
5. To understand the importance of the twelve principles of green chemistry.
6. Explore the modern waste segregation techniques to reduce the demand for virgin resources.

11.1 Introduction

Green Manufacturing refers to the practice of using energy-efficient technologies, appliances, and renewable energy resources, and the reduction of harmful emissions, to have the sustainable industrial production generating a minimum carbon footprint. On the other hand, waste management refers to reducing, reusing, recycling, repurposing, and repairing or refurbishing waste to implement the circular economy. Together, green manufacturing and waste management lead to long-term ecological sustainability along with compliance with environmental regulations. Traditional production methods and waste disposal methods create pollution and depletion of natural resources.

In this chapter, we shall explore various green practices to integrate the manufacturing process and efficient waste management for a sustainable resource-conserving industrial future.

11.2 Industrial Energy Efficiency Practices

Industries are the major consumers of energy. Increasing energy efficiency in the industrial sector reduces the cost of production and improves sustainability. According to the International Energy

Agency (IEA), more than 40% of global energy is consumed in the industrial sector. Cement, Steel, mining, and chemical industries are major energy-consuming industries. Since a huge part of the energy generated is used in industries, they are also responsible for significant greenhouse gas emissions and carbon footprint. It is especially important to adopt energy-efficient practices in the industrial sector to reduce climate change impacts and meet global sustainability goals. These practices may include advanced machinery or optimization techniques in the existing network. Let us explore these practices in detail.

- Energy-efficient equipment: Energy efficiency can be drastically increased by replacing the old machinery with new, more advanced, and energy-efficient machines. It may include the latest electrical drives, LED lighting, and efficient HVAC systems. Variable Frequency Drives (VFDs) are extremely helpful in reducing energy consumption by automatically adjusting the motor speed to meet load demand. Looking towards the trend of global warming and ever-increasing atmospheric temperature, the importance of HVAC systems is increasing. Smart temperature control and superior quality insulation can save a significant amount of energy.
- Process Optimization: It refers to the reduction in energy consumption and waste during manufacturing, transportation, and storage. It involves lean manufacturing (that is, waste reduction and eliminating non-productive activities), Just-in-Time (JIT) production, and judicious use of resources.
- Heat Recovery Systems: Heat is generated in many industries as a by-product. Rather than wasting this heat, it is utilized in other purposeful activities. In a cogeneration power plant, electricity is generated using the waste heat from another process. This electricity can then be used to fulfil the needs of industry. Heat exchangers and heat pumps

can also be used to retrieve heat for utilization in other useful activities.

- In many industries, like cement and metal, a significant part of energy is wasted as heat along with flue gas. This heat can also be captured to preheat the combustion air.
- In the same way, those industries that use a large amount of water for cooling can also recover heat from wastewater and use it for heating purposes.

11.3 Energy Management System (EnMS)

An energy management System helps industries to monitor and control energy usage by providing a systematic, structured approach to improve energy efficiency. The ISO 50001 is the well-known standard for this purpose. Key components of EnMS are Energy planning, energy performance, and continuous improvement. Key benefits of EnMS are cost savings, improved sustainability, and compliance with regulations. Now, let us consider a few case studies highlighting the energy efficiency measures. It may involve Green Supply Chain Management (GSCM) for sustainable procurement, supplier collaboration, and logistics optimization to enhance energy efficiency. Water efficiency can be increased by water recycling, water reuse, reverse osmosis, and waterless processing to reduce water consumption. 3D printing can be used to repurpose or recycle the material to reduce the impact on the environment. IoT can be integrated with the manufacturing operations to provide real-time insights and reduce waste generation and downtime through predictive maintenance. AI and ML can also be used to improve quality control and process optimization.

Toyota and many other auto manufacturing companies have high energy demands for welding, painting, and assembling *(Toyota Sustainability Report, 2023)*. These companies have implemented robots, LED lighting, and improved algorithms to minimize energy consumption. ArcelorMittal has implemented the advanced heat recovery system in its steel plants to recover waste heat from blast furnaces *(ArcelorMittal Corporate Responsibility Report, 2022)*. By utilizing this waste heat, they have reduced dependence on fossil fuels and lowered their carbon footprint. The main challenges lying on the path of industrial energy efficiency are the high initial investment required for purchasing the energy-efficient equipment, the dearth of experts, the lack of awareness among the industrialists, and outdated energy regulations. Despite these challenges, industrial energy efficiency is the need of the hour and has to be implemented. Emerging technologies like AI, IoT, and Machine Learning (ML) may enhance energy efficiency shortly through real-time monitoring and integration of renewable energy sources.

11.4 Energy Audit

It is an evaluation to find out the scope of improving energy efficiency, reducing waste, and lowering the cost of production. An energy audit provides a comprehensive analysis to identify the locations of energy waste in any process. It also recommends ways to improve efficiency. It also helps an enterprise to track energy usage, energy demand, and patterns of load to predict the future scope of energy efficiency. The main goal of an energy audit is to suggest cost-effective measures for energy savings, waste reduction, and decreasing greenhouse gas emissions to lead a sustainable development. It also helps in the reduction of electricity bills. Energy audits can be classified into the following two types depending on the required detail and complexity:

1. Preliminary Energy Audit: This is preliminary, so it is also known as a walkthrough audit. A comprehensive, in-depth data collection and analysis is not available in this case. Major energy-consuming processes and sites are examined to find the scope of energy efficiency. It is a quick and low-cost assessment.
2. Detailed Energy Audit: In this type of audit, all equipment is evaluated for the performance assessment through energy meters. It also includes lighting, HVAC, motors, and other machinery. Thermal performance of buildings and insulation is also carried out to seal the energy leakage. Detailed

analysis is carried out based on comprehensive data collection. Thereafter, suitable measures are suggested to improve the energy efficiency. A detailed energy audit is often comprehensive, costly, and requires more time.

Stages of Energy Audit:
The following are the various stages of an energy audit:

Figure 11.1: Steps of Energy Audit

1. Initial Assessment: This stage includes the data collection related to energy consumption, operating hours, location, climate, number of employees, existing practices, and other parameters that can impact the energy consumption. A primary analysis is done to review and identify the energy consumption patterns and possible losses in major energy consumption points.
2. On-site Assessment: It includes visual inspection, inspection of lighting, HVAC areas, insulation, motors, drives, and mechanical energy transmission structures. This helps the energy auditor to identify the inefficiencies. If required, the auditor may install energy meters and sensors to measure and track the real-time parameters. Major equipment like compressors, boilers, industrial ovens, and central conditioning systems is evaluated for inefficiencies.
3. Data Analysis: Based on the data collected in the above steps, now auditors analyse the data to understand the energy consumption pattern, peak energy usage periods, brainstorm alternative ways to divide the peak load, and find out the improvement in process, replacement of equipment, reduction of waste energy, and poor insulation.

4. Reporting: The last step of an energy audit is to submit the report. It includes a summary of findings and suggestions for improving energy efficiency. It also includes the estimated energy savings and payback period for suggested improvements. Feedback is also provided to improve the overall efficiency of a plant.

Benefits of Energy Audit: An Energy audit offers many advantages to a building, business, industry, or commercial hub. The most significant advantage is cost reduction. The company or an organization can reduce the electricity bill by implementing the recommendations of the energy audit team. An energy audit also directly helps in the reduction of greenhouse gas emissions to improve sustainability and mitigate the ill effects of climate change. An energy audit helps in compliance with regulations to avoid penalties. For example, a simple suggestion of installing a capacitor bank at the meter room can improve the power factor of electricity, reduce the electricity bills, and avoid the penalty for power factor. Many times, we are not able to point out our drawbacks because we are habituated to ourselves, but a third-party inspection, like an energy audit, can see those loopholes and seal those leaks to improve operational efficiencies, leading to enhanced productivity. Energy audit improves sustainability performance, leading to an upliftment of the company's image in the market. This can attract the environment-conscious investors and customers.

Challenges:

No doubt that energy audits are essential and helpful to establish sustainable practices, reduce utility bills, and save our environment, but there are certain challenges in implementing energy audits. The most significant challenge is initial investment, as it requires the upgradation of old machinery, the replacement of existing lighting with LEDs, etc. People are always reluctant to accept change, so sometimes it becomes difficult to convince the owner of the company to undergo an energy audit, even when it is helpful for them in the long term. A lack of awareness about energy audits and their benefits, as well as a lack of responsibility towards the environment, are also major factors in the implementation of energy audits. Small, micro, and medium-sized businesses sometimes can not afford the cost of an energy audit and required upgradation.

To provide additional ideas about energy audits, the following is a sample energy audit report, which may apply to a small industry.

SAMPLE ENERGY AUDIT REPORT

Title: Energy Audit Report of Hindustan Textile Solutions Ltd, Surat, India
Location: Plot No. 257D, Kim GIDC, Surat, Gujarat, India
Industry Type: Textile Processing Unit
Audit Type: Preliminary Walkthrough Energy Audit
Date of Audit: 28-30 June 2025
Auditor: Global GreenTech Consultants LLP, Gandhinagar, Gujarat, India

****

1. Objective: To assess the existing machinery in a plant for current energy consumption, to identify the inefficiencies, and recommend energy-saving opportunities with a payback period.
2. Plant Profile:
    - Installed Load: 75 kW
    - Operating hours: 12 hours per day, 6 days per week
    - Major Equipment:
        - Air compressor
        - Jet dying machine
        - Saree cleaning and drying machine
        - Steam boiler
        - Lighting
        - Motorized washing units
        -
3. Energy Consumption Summary for the last year:

Source	Annual Consumption	Cost in INR
Electricity	75,000 kWh	6,85,000
Diesel for the boiler	9500 litters	8,55,000
Total Energy Expenses		15,40,000

4. Major Findings:
   A. Electrical:
   - Inefficient and old lighting system, outdated power-consuming CFLs, and Tube lights.
   - Most of the motors are operating at 60-75% load, yet without variable frequency drives (VFDs).
   - Power factor: 0.81 lagging

   B. Thermal:
   - Boiler operation at ~56% , flue gas heat not fully utilized
   - Heat loss due to uninsulated /torn insulation (~15 %)
   - Heat waste in hot slurry

5. Energy Saving Opportunities (ESOs):

Sr. No.	Recommendation	Approx. Investment	Annual Savings	Payback period
1	Replace CFLs with LED lighting	₹28,000	₹15,000	1.6 year
2	Install VFDs on the main three motors	₹1,30,000	₹90,000	1.5 year
3	Insulation improvement for boiler and steam pipes	₹ 65,000	₹ 52,000	0.9 year
4	Installation of a capacitor bank for power factor correction	₹25,000	₹25,000	1 year
5	Recover heat from flue gas and slurry	₹ 90,000	₹ 44,000	1.8 years

6. Recommendations:
   - Switch to solar water heater or hybrid energy if feasible

- Conduct a detailed energy audit every year.
- Train operation and maintenance staff.
- Use natural daylight, whitewash the walls, and replace old fans.
- Replace old ACS with inverter ACs

7. Conclusion: By implementing the above recommendations, Hindustan Textile Solutions Ltd, Surat can save approximately up to ₹ 2.26 lakh annually (which is approx. 15% of total energy cost) with an average payback period of ~1.4 years. This may benefit the company not just economically but also in achieving the climate goals.

==

11.5 Green Chemistry

Green chemistry refers to those chemical processes that are environmentally friendly, economically viable, and energy efficient to produce. Thus, green chemistry also helps to reduce the poisonous and hazardous wastes during and after the product's life. Thus, it offers products with minimum environmental impact, minimum waste, longer-life products with minimum energy consumption, so it is also known as sustainable chemistry. Paul Anastas and John Warner framed **twelve principles of green chemistry** in 1998 to guide researchers, engineers, and industrialists to create safe and sustainable chemical production. They are:

1. Prevention: Let the waste be prevented from being created.
2. Atom Economy: Design a method to maximise the inclusion of all raw materials in the process to get a final product, so that minimum waste is produced.
3. Less Hazardous Chemical Synthesis
4. Designing Safer Chemicals
5. Safer Solvents and Chemicals
6. Design for Energy Efficiency
7. Use of Renewables
8. Reduce Derivatives

9. Catalysis: Use those catalysts that are required in tiny amounts and generate less waste.
10. Design for Degradation: The Product should be designed in such a way that it can be decomposed into small non-hazardous components.
11. Pollution Prevention
12. Safer Chemistry for Accident Prevention

Importance of Green Chemistry

The majority of products we use today are manufactured in chemical industries, so chemical industries are consuming a massive portion of energy and resources. This sector directly or indirectly plays a vital role in many other sectors like medicines, agriculture, plastics, cement, paint, and manufacturing. In many processes, hazardous and/or toxic wastes are produced, which are harmful to human, animal, and plant health. It pollutes our air, water, and soil. Green chemistry can help reduce these wastes and have an impact on the environment. It can be achieved by:

- Waste Reduction: In the case of production through the traditional chemical process, a large amount of hazardous waste is produced. Green chemistry focuses on waste reduction. By-products of one process may be used as raw material in another process.
- Energy Efficiency: As said earlier, chemical process consumes a large amount of energy. The traditional production method requires a large amount of energy, while green chemistry focuses on processes that require minimal energy.
- Safe Chemicals: While traditional chemical processes do not care for the toxicity of solvents and chemicals used, green chemistry prioritizes the health of workers and the environment. Alternative safe chemicals are preferred for the same product.
- Sustainability: Traditional chemical industries rely on conventional fossil fuels for energy and petroleum-like

resources, while green chemistry promotes the use of renewable sources for energy and promotes the use of renewable materials like plant-based feedstocks as raw materials. For example, bioethanol, biodiesel, and biogas are alternatives to fossil fuels. Bioplastics, which are also produced from corn starch, are an alternative to petroleum-based products.

- Closed loop cycle: Green chemistry promotes the closed loop system involving the 5R principle of sustainability to reduce the use of new raw materials and utilize the waste.
- Catalysis: You might be able to recall the "catalyst" from your science chapter in class 10th, which taught us that catalysts are the substances that speed up the reactions, without being consumed in the process. It helps a higher reaction rate, reduced energy input, with minimal by-products. Enzymes are special catalysts that can function even at moderate temperatures and pressures, reducing energy consumption. Our digestive process heavily depends upon these enzymes for breaking down and digesting the food very quickly. You might wonder, but digestion also involves several chemical processes.

Challenges for Green Chemistry

The very first challenge for the implementation of Green Technology is the economic viability, because in some cases it may require high investments and operational costs. It may be expensive and complex for the existing industries already working at a mass level. Many technological barriers are also present, as green chemistry is still in the development stage. The transition from the lab to industry may take decades. Existing frameworks and regulations may prove hurdles in implementing green chemistry. At last, the ultimate consumers, who are the public, will need an awareness to be aware of and accept costly, but environmentally friendly products produced through green chemistry to understand their long-term advantages. Despite these challenges, many companies are putting in vigorous efforts into green chemistry. For example, BASF has developed many eco-friendly chemicals and processes, as well as biodegradable agrochemicals. Many pharmaceutical companies, including Pfizer, are focusing on biocatalysts in drug synthesis to get rid of toxic solvents and hazardous wastes.

11.6 Industrial Ecology and Circular Economy

These two concepts are closely interlinked with each other. They focus on sustainable production systems, encourage the reuse of waste, adopt more energy-efficient processes, and conserve resources. Industrial ecology focuses on utilizing waste from one process as raw material for another purpose; on the other hand, the circular economy focuses on a closed loop. Let us explore these two terms in detail in the following section.

Industrial Ecology: Industrial ecology tries to replicate the concepts of the natural ecosystem. It includes designing closed-loop processes, where waste or by-product of one process is utilized as raw material for another process to get another useful product. In this case, raw materials, energy, and waste are considered part of the industrial ecosystem. Key components of industrial ecology are material flow analysis, life cycle assessment of products, eco-industrial parks, and industrial symbiosis. The key role of industrial ecology is resource optimization, waste minimization, energy efficiency, waste-to-value, and product life cycle management.

Circular Economy: The Traditional linear economy system follows the rule of “take, make, and dispose,” while the circular economy system follows the rule of “reduce, reuse, and recycle.” A traditional linear ecosystem does not care about the wastage of natural resources; its sole aim is to earn profit. Thus, the environment is exploited a lot. The circular economy, on the other hand, focuses on the conservation of natural resources by waste reduction, designing longer-life products, and recovering or reusing the product after its useful life. The circular economy stands on three principles, viz., long-life design, extension of product life, and recovery of the material at the end of product life.

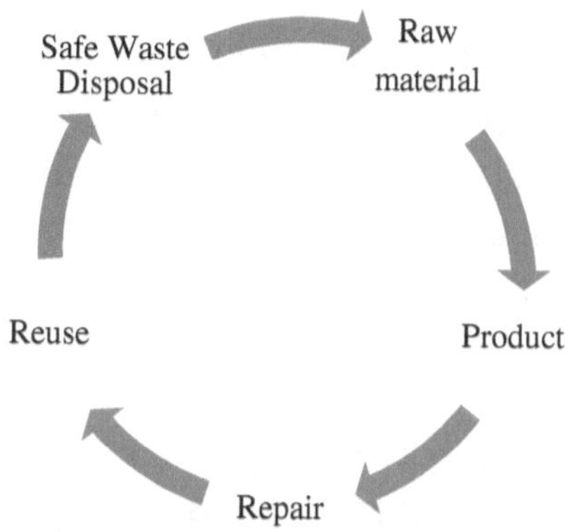

Figure 11.2 Ideal Product Cycle: From Raw material to Raw material

Key principles of the circular economy are:

1. Design for Disassembly: This principle of circular economy emphasises designing the products in parts, so that these components can be reused, repurposed, refurbished, or recycled after their useful life. This type of product design reduces waste.
2. Products as a Service (PaaS): Rather than selling the product, offering it for rent as a service is always beneficial environmentally. For example, instead of selling a car, a company may provide it as a service. The ownership of the car shall always remain with the company. In this case, a car or any other product will get professional regular service, and will have better options for reusability and recycling. One more advantage of this system is that a smaller number of cars will be required, as people will rent a car only when it is required. So, the car will not lie idle in the people's parking.

3. Closed-Loop Manufacturing: As the name itself suggests, in this case, waste materials are reused in the production of novel items. Many companies are promoting buy-back schemes or exchange schemes to achieve the target of closed-loop manufacturing.
4. Recycling and Upcycling: Other ways are being explored to utilize the waste into useful products.
5. Modular Design: If any product is manufactured in modular design, then only the faulty part needs to be replaced rather than replacing the whole product, thus increasing the product's life and reducing waste.

Understanding the importance of the circular economy, many companies have now started to focus on this area. *Interface* is a leading manufacturer of carpets that collects discarded fishing nets from coastal areas, processes them to fit as raw material, and uses them to manufacture new carpets to reduce dependence on virgin resources. This initiative has contributed to marine conservation, reduced waste, and ecological sustainability. Another notable initiative is taken by the leading car manufacturing company from France, *Renault*. It has initiated the Renault Re-Factory program, where it focuses on reusing and remanufacturing new parts from old materials. Important parts like engine, battery, etc, are retrieved from end-of-life cars and are refurbished to fit in new cars. This method reduces waste as well as production costs.

Under India's 2023 G20 presidency, 39 companies from 11 countries, like Tata, Maruti, JSW, Coca-Cola, Unilever, Nestle, agreed to a circular economy model under the Resource Efficiency and Circular Economy Industry Coalition (RECEIC). G20 environment ministers express commitment to the industry-led EPR implementation.

***Thinking Corner**:*

You are on an evening walk with your friends in a market. You saw sweet mulberries at a shop and purchased 250 grams to enjoy with friends. Shopkeeper packed mulberries in a plastic bag. What will be your stand (a) to accept in polyethene bag (b) to encourage shopkeeper to give in paper/cloth bag (c) not to buy mulberries in spite of watering mouth!

11.7 Waste Segregation and Upcycling Technologies

In the era of industrialization, mass production has led to a pile of products in the market. Various governments across the world are also encouraging the industrial sector to produce more to support growth and employment. Easy financing and quicker loans have motivated people to purchase consumer goods at a record level. This vicious cycle has resulted in mountains of waste in almost all cities. Waste segregation refers to the process of separating diverse types of waste for proper recycling and minimizing the contamination of water, air, and land. Broadly, waste can be classified into two categories. Municipal waste and Industrial waste. Municipal waste may contain food remains, paper, plastic, old consumer goods, wood scraps, bio-waste, etc. While industrial waste is more diverse, ranging from hazardous solvents, paints, chemicals, hot slurries, ash, metallic scraps, etc. Proper segregation of both these wastes is essential to conserve our resources. Common people, shop owners, commercial complex owners, and industrialists should be encouraged to sort their waste into different bins for paper, organic waste, e-waste, and hazardous materials. Many countries have adopted smart waste bins using IoT to auto-separate the waste based on its content for easier recycling. India has also initiated the segregation of wet, dry, and e-waste at the source level. The following are a few waste segregation technologies:

1. Optical Sorting: This system is widely used in recycling complexes for precise and fast waste segregation for paper, metal, and plastics. It is composed of cameras and light

sensors to analyse the waste depending on the colour, shape, and material properties. When this waste passes along a conveyor, this system identifies the waste and separates it into different bins according to its category.

2. Electrostatic Separation: This method is used when waste is a mix of metal and nonmetals. Optical sorting method and traditional waste segregation methods can not help in this case. This method uses the electrical properties of waste material to separate it into metals and nonmetals, such as plastic or wood, based on their electrical charge.
3. X-Ray Fluorescence (XRF): This method is useful in the separation of metals, plastics, and minerals in the waste stream. This technique separates the material based on its elemental composition. It also helps to identify the hazardous materials to ensure non-poisonous waste for recycling.
4. Biological Waste Sorting: This technique uses the natural ability of microbes to break down organic waste like food waste, agricultural residues, and other biodegradable waste. This method can separate organic and inorganic waste very efficiently and cost-effectively, but it is time-consuming.
5. Robotic Sorting: This is the fastest but costliest method. It involves the use of AI-powered robots to identify and separate the waste. For this, these robots are equipped with specific algorithms to identify the specific type of material and sort it into a particular bin.

The importance of Upcycling:

Upcycling refers to the process of converting a waste material into a useful product. Upcycling is different from recycling. In recycling, waste is broken down into small pieces to form raw material to prepare a new product. While in upcycling, the waste product is directly converted into a new, useful product without much processing. Thus, upcycling not just helps in waste reduction but also enhances the value of a product. Its importance is mentioned below:

1. Waste reduction: It gives waste a second life.
2. Conservation of resources: Use of virgin resources is reduced. So, the environmental impact is reduced.
3. Economic opportunities: New jobs may be created in the design, manufacturing, and retail sectors.
4. Innovative product development: It encourages creativity and innovation to design a quality, useful product from waste. *Have you ever been given a project to create a "Best from Waste" in your school? Recall it. Now, brainstorm how you can replicate today!*

***Activity**: You might have seen kabbadiwalas (scrap collectors) and ragpickers collecting waste papers, metal-plastic scrap, broken-unrepairable-obsolete gadgets, and other reusable items.*

(1) List different types of materials you and your family members typically use:

(a) Throws in the garbage

(b) sells/donates to kabbadiwalas

*(2) Estimate the approximate **weight** of waste generated by your family in a month. It may include everyday items such as old toothbrushes, newspapers, old gadgets, textbooks, stationery, worn-out footwear (chappals), milk pouches, plastic carry bags, toys, discarded furniture/ mattresses, cosmetic containers, bottles/ boxes for groceries, wrappers, and more.*

Now, try to imagine the carbon footprint per person for your family. Consider how much of this waste can be recycled or reused.

Be respectful and empathetic towards kabbadiwalas and ragpickers, as they are the unsung heroes of the circular economy, to play a crucial role in promoting sustainability. If possible, donate old items and scrap to them – not just to reduce waste, but to support the livelihoods of those, are not working for luxury but to meet the basic needs like food and shelter.

Before ending the discussion on upcycling, the case study of BMW's Closed Loop System is worth discussing. BMW, a leading car manufacturing company, has implemented a closed-loop upcycling system in its group companies. Advanced sorting mechanisms are used to reclaim materials like metal sheets, steel, and plastic from old cars and are upcycled into new cars. By following such practices, all companies can reduce waste, save energy, and conserve natural resources. *(BMW Group Sustainability Report, 2023).*

11.8 E-Waste Management

Remember a day without any electronic gadgets like a mobile, a laptop, or a computer in your life! It is pretty hard, isn't it? In recent decades, the production and consumption of electronic gadgets have accelerated at rocket speed! At the same time, this development has led to unforeseen calamity of piles of electronic waste, commonly referred to as E-Waste. E-Waste includes all sorts of used, discarded, or outdated electronic gadgets like smartphones, computers, televisions, refrigerators, AC, household appliances, and many other commercial or industrial equipment that are composed of hazardous materials like mercury, lead, and others

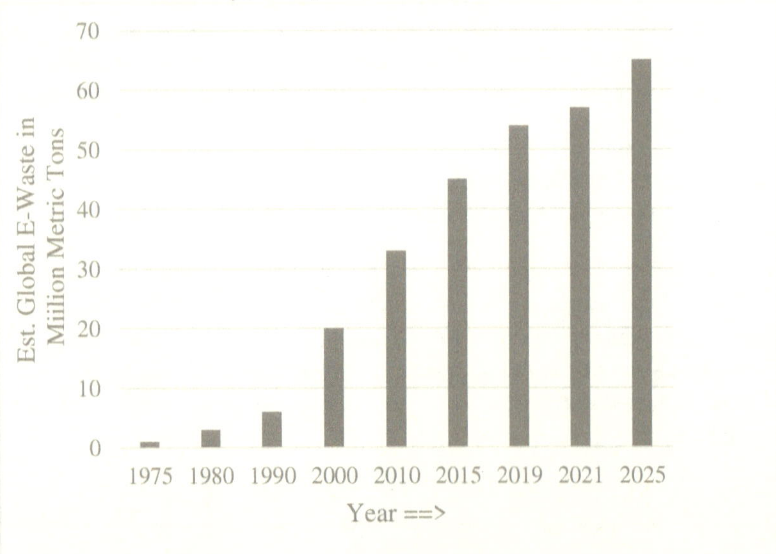

Figure 11.3 Global E-waste Generation in Million Metric Tonnes (Data sourced from United Nations Global E-waste Monitor reports

Figure 11.3 shows the exponential rising trend of e-waste globally, while Figure 11.4 helps to understand the stake of various categories producing e-waste.

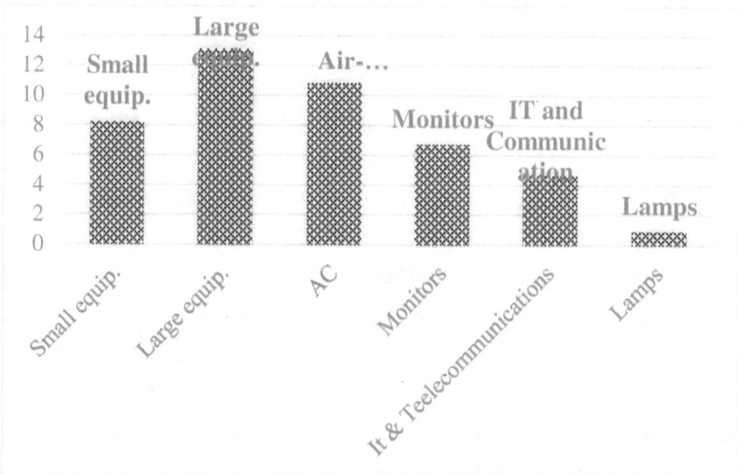

Figure 11.4 A Chart to represent the share of various e-wastes. Data Source: Global E-waste Monitor Report, 2020

Impacts of E-Waste: Most of the electronic devices contain harmful materials and minerals like lead, mercury, cadmium, etc., which pollute our air, water, and soil if they are not properly disposed of. If e-waste is dumped into landfills or disposed of in common garbage, then it can penetrate the soil and pollute our groundwater. If e-waste is burned, then it may release toxic fumes, which are harmful to our health. Prolonged exposure to such toxic gases may cause serious health problems and diseases. During the production of electronic gadgets, precious metals as well as rare Earth elements are used; if these gadgets are not properly recycled, then these precious minerals are lost, mandating new resource excavation.

E-waste Management Methods

It is especially important to bring back this e-waste to reuse and recycle, and recover the precious minerals and extract hazardous substances. Many nations and companies have initiated a few strategies for the maximum recovery of sold products.

1. Take-Back Schemes: In this case, the consumers are motivated to return the old products for proper disposal and recycling. This helps to divert the e-waste from being dropped into ponds, water bodies, and other landfills. Dedicated collection centres are established where consumers can return the old, unused, broken, or working but unwanted, obsolete, and outdated products. Customers may be offered a replacement scheme or a lump sum of money in exchange for old products.
2. Certified E-Waste Recycling centres may help in the safe handling, transportation, and disposal of e-waste. Their work includes de-manufacturing, separating valuable materials, and separating toxic materials.
3. Safe disposal: Under expert supervision, all components and minerals are disposed of after recovery of precious and rare-earth minerals. Controlled incineration of certain e-waste is carried out to dispose of nonrecyclable components. Safe landfilling is proposed only for nonrecyclable components.

E-Waste Recycling Technologies:

Some of the advanced e-waste recycling/ segregation technologies are explained below:

1. Mechanical Separation: In this method, shredding machines are used to break the e-waste into smaller pieces. Then these smaller pieces are segregated into dissimilar categories like plastics, metals, glass, and others. Different strategies like vibration, magnetic separation, air blowing, and sieving may be used to separate dissimilar materials. This is the easiest, traditional, and widely used technique to separate any type of waste, including e-waste.
2. Hydrometallurgical Process: This method is used to separate precious metals like gold, silver, copper, and rare-earth

minerals using aqueous solutions. This method is more environmentally friendly than the traditional smelting process.

3. Pyrometallurgical Process: This method is energy-consuming because it uses high temperatures to extract precious metals like gold and palladium from printed circuit boards. This process should be carried out in a controlled environment to avoid harmful gas emissions.
4. Biotechnological Recycling: In this method, microorganisms are employed to separate metals from e-waste. This process is also environmentally friendly because harmful chemical solvents are not used, and it is also energy-efficient because no high temperature is required in this case. Enzyme-based recycling can also be used to separate plastic from e-waste.
5. AI-Based Sorting: In this method, AI-based technology is used to separate the e-waste using machine learning, image processing, and a robotics arm. This method is fast, accurate, and reduces human exposure to harmful material, but it is costly.

> ***Thinking Corner**: You might have reached the age of 18. By this time, how many mobile phones have you used? How many of them have been recycled? Did you ever throw your old phones in a lake, dustbin, or garbage? Do you know it can poison our water and food?*

Extended Producer Responsibility (EPR): EPR programs mandate that manufacturers be responsible for the product manufactured during its useful life and the safe disposal of the product after its life. This ensures better e-waste management professionally. The manufacturers will be prompted to design a product in such a way that the product requires minimum maintenance, easy repair, and safe disposal. Many countries have implemented the EPR laws.

For example, in the European Union, the Waste Electrical and Electronics Equipment Directive (WEEE) framework has been established, while in India, the E-Waste (Management) Rules are framed, which track the e-waste and ensure its maximum recycling. The United Nations has also initiated the E-Waste Coalition to promote a sustainable e-waste management system. Dell, a global leader in the manufacturing of electronics, has also initiated the Take-Back program, under which the company aims to recycle 100% of its products.

What can we do?

What is our duty in this regard? It is not just the duty of governments!

Giant companies, governments across the globe, United Nation, and many NGOs are putting their efforts, but being a responsible HUMAN, we should also pledge to dispose of our smart phones, washing machines, AC, cooler, Computer, Laptops, and other electronics gadgets in a safer way by depositing the broken, outdated, and non-repairable items in a concern store for proper recycling. Think before you throw.

-: EXERCISE:-

Section A: Short Answer Questions

1. Define Green Manufacturing.
2. Mention two goals of green manufacturing.
3. State the 5R principle in waste management.
4. List eco-friendly products that are used in green manufacturing.
5. Define industrial symbiosis.
6. State any three sources of industrial waste.
7. Define e-waste.
8. List major e-waste constituents.
9. State the difference between pre-consumer and post-consumer waste.
10. Explain zero-waste production.

Section B: Long Answer Questions

1. Explain how green manufacturing supports sustainable industrial development.
2. Discuss the social and environmental advantages of adopting green manufacturing.
3. Explain how LCA can help in reducing the environmental impact of products.
4. Provide strategies to achieve zero-waste manufacturing in industries.
5. Explain different types of industrial waste. Suggest treatments for their safe disposal.
6. Discuss the challenges and opportunities for the implementation of green manufacturing in developing countries like India.
7. Suggest steps to reduce the environmental impact of e-waste.
8. Suggest strategies for the safe recycling of e-waste.
9. Evaluate the role of digital technologies in green manufacturing to reduce the carbon footprint.
10. Suppose you are the CEO of a Smartphone manufacturing company. Propose a waste management framework for green manufacturing and waste reduction.
11. Explain how traditional manufacturing is less sustainable than green manufacturing.
12. Discuss the challenges of adopting Green Technology.

Section C: Multiple Choice Q and A

1. The primary focus of Green manufacturing is:
   a) Maximizing fossil fuel use
   b) Reducing carbon footprint and using renewables
   c) Increasing linear production
   d) Ignoring waste management.
2. ……… % of global power is consumed by industry.?
   a) Over 30% b) Over 40% c) 50% d) 60%
3. Variable Frequency Drives (VFDs) save energy by:
   a) Running motors at a constant high speed
   b) Adjusting motor speed to load demand
   c) Increasing heat waste
   d) Decreasing speed.

4. Heat recovery systems in industries are used to:
   a) Discard excess heat
   b) Preheat combustion air or other processes
   c) Increase flue gas emissions
   d) For employee safety
5. ISO 50001 is the standard for:
   a) Waste disposal
   b) Energy Management Systems (EnMS)
   c) Quality management
   d) Chemical safety.
6. A preliminary energy audit is also called a:
   a) Detailed audit b) Walkthrough audit
   c) Thermal audit d) Financial audit.
7. On-site assessment in an energy audit includes:
   a) Only reviewing electricity bills
   b) Inspecting HVAC, insulation, and motors
   c) Analysing marketing data
   d) Ignoring equipment performance.
8. True/False: Detailed energy audits are cheaper than preliminary audits.

9. An energy audit helps .... the electricity __________ bills.

10. True/False: Energy audits are unnecessary for compliance with regulations.

11. The twelve principles of green chemistry were formulated by:
    a) Einstein and Bohr
    b) Paul Anastas and John Warner
    c) John Dalton and Chadwick
    d) Marie Curie
12. Atom economy emphasizes:
    a) Maximizing waste
    b) Incorporating all raw materials into the final product
    c) Using toxic solvents
    d) Ignoring catalysis to promote sustainability.

13. True/False: Green chemistry uses twelve basic principles for sustainability.

14. __________ increases the rate of reactions.

15. A challenge in green chemistry is:
    a) Low public awareness
    b) Economic viability
    c) Simplified lab-to-industry transition
    d) Excessive government support.
16. Industrial ecology mimics:
    a) Linear production b) Natural ecosystems
    c) Competitive markets, d) Fossil fuel dependence
17. True/False: A circular economy follows the "take-make-dispose" model.

18. *Interface* upcycles discarded __________ nets to make carpets.

19. "Products as a Service" (PaaS) involves:
    a) Selling disposable goods
    b) Leasing products instead of selling ownership
    c) Increasing e-waste
    d) Ignoring recycling.
20. Design for disassembly aims to:
    a) Increase landfill waste
    b) Facilitate the reuse/recycling of product parts
    c) Shorten product lifespan
    d) Use non-recyclable materials.
21. True/False: E-waste includes only smartphones and laptops.

22. High temperature is used to extract metals from e-waste in the __________ process.

23. In Extended Producer Responsibility (EPR)*:
    a) Consumers handle all e-waste
    b) Manufacturers manage product end-of-life disposal

c) Open burning of e-waste
d) No recycling requirements

24. AI-based sorting in e-waste is:
    a) Only for organic waste
    b) Fast, accurate, and costly
    c) Slow but cheap
    d) Illegal in most countries.
25. True/False: Hydrometallurgical processes are more eco-friendly than smelting.

26. Upcycling differs from recycling because it:
    a) Breaks waste into raw materials
    b) Repurposes waste directly without breakdown
    c) Can not be applied to printers
    d) Only works for plastics.
27. BMW's closed-loop system upcycles materials from old …….

28. True/False: Optical sorting uses sound waves to segregate waste.

29. X-Ray Fluorescence (XRF) separates waste based on:
    a) Colour b) Elemental composition
    c) Smell d) Weight.
30. The WEEE Directive is a __________ program for e-waste.

Answers Keys:

1.B, 2.B, 3.B, 4.B, 5.B, 6.B, 7.B, 8. False, 9. reduce, 10. False 11. B, 12. B, 13. False, 14. catalyst, 15. B, 16. B, 17. False, 18. Fishing, 19. B, 20. B, 21. False, 22. Pyrometallurgical, 23. B*, 24. B, 25. True 26. B, 27. Cars, 28. False, 29. B, 30. EU's EPR

**In India, EPR is governed by E-Waste (management) rules*

संगच्छध्वं सं वदध्वं सं वो मनांसि जानताम्।
देवा भागं यथा पूर्वे सञ्जानाना उपासते॥

(ऋग्वेद 10.191.2)

"Let us all move together, let our minds be in agreement, let us speak in harmony to be united for peace and prosperity."

(Rigveda 10.191.2)

Chapter 12

GLOBAL CONTEXT AND INTERNATIONAL POLICIES

Learning Outcomes:

After reading this chapter, the learner will be able to:

1. Discuss the role and importance of global environmental treaties.
2. Discuss the role of international organizations like UNFCC, IPCC, and UNEP
3. Explore the effectiveness of global and regional treaties and funding mechanisms to mitigate climate change and adopt sustainability.
4. Analyse the role and responsibility of governments, businesses, industries, and common people towards sustainability.

12.1 Introduction

In the modern era of technology, the world has become a global village. No nation can remain unaffected by the actions of other countries. Air, water, and land ecology are collectively connected by the biosphere. Rapid industrialization in the past century has brought many fruitful results for us, but at the same time, it has also brought a calamity of climate change due to excessive exploitation of natural resources and heavy pollution. No community, region, or nation alone can fight against this menace of climate change. Collective and coordinated actions are required from all the stakeholders, like people, organisations, and countries. International policies are required to motivate and guide the stakeholders to implement the regulations and frameworks to ensure a safer future.

Real Inspiration

Rehan, living in a village near Varanasi, built a solar lantern with the help of broken toys and e-waste after watching videos in cyber café. This solar lantern is helpful for her weaver mother to work more to fulfil the dream of her son to become a doctor.

Source: The Better India, 2022

Understanding the immediate need for action to combat climate change and its ill effects, many nations have started to take affirmative steps. The Paris Agreement (2015) and the Kyoto Protocol (1997) are milestones in the direction of putting efforts to reduce global temperature rise and net-zero carbon emissions. The United Nations Framework Convention on Climate Change (UNFCCC), International Panel on Climate Change (IPCC), and United Nations Environment Programme (UNEP) are a few other organisations that are actively involved in promoting steps to reduce global temperature, pollution, and save the global biosphere for future generations. These institutions also act as a bridge between developed and developing countries for technology transfer, financial support, and promoting equality. Many regional and bilateral efforts are playing a key role in global Green Technology. European Green Deal, Nationally Determined Contributions (NDCs) of India, and Renewable Energy Investments from China reflect awareness at a local level to contribute at the global level. Multinational funding assistance, multinational research foundations, and the international carbon market show the Interconnectedness of the novel green ecosystem.

This chapter explores a few frameworks in the light of Green Technology and sustainability, and their role and impact at the global level.

> ***Thinking Corner: Carbon Credits or Morals Debits?***
> *A rich country offsets its pollution by funding forests in a poor country (deforested by the rich country for minerals).*
> *What do you think, are carbon markets solving the climate change issue or shifting the burden smartly?*

12.2 International Treaties

The Paris Agreement

This was the first comprehensive global landmark treaty to combat climate change. It was adopted in 2015 (at COP21, Paris, France) through the United Nations Framework Convention on Climate Change (UNFCCC) [15]. It was signed by almost all countries of the world, showing a voluntary commitment to combat climate change. The core intention of the Paris Agreement is to hold the global warming well below 2°C with an intention to limit the global temperature rise below 1.5 °C above the pre-industrial level [15]. This 1.5°C target was coined by the Intergovernmental Panel on Climate Change (IPCC), expressing a fear that crossing the threshold of 1.5 °C could lead to an irreversible cycle of extreme weather events. Understanding the need for immediate steps, all participating member countries have been planning Nationally Determined Contributions (NDCs) for the reduction of greenhouse gas emissions. The Paris Agreement applies to both developed and developing countries with common but differentiated responsibilities.

A standardized Enhanced Transparency Framework (ETF) has also been drafted to report the annual greenhouse gas emissions and achieved progress, to bring transparency and accountability. The Ratchet Mechanism is the unique feature of the Paris Agreement, which mandates the member countries to review and update their NDCs to a higher level to ensure progress over time. The Paris Agreement also takes care of climate finance, capacity building, and climate adaptation. Financial assistance is provided by developed countries to low-income countries through the Green Climate Fund (GCF).

Active efforts of the Paris Agreement have initiated many activities to mitigate climate change worldwide. Many countries pledged to have net-zero emissions by mid-century through their NDCs. Many countries, like the United Kingdom, South Korea, Japan, India, and the EU, have integrated renewable energy policies to align and achieve the targets of NDCs. Despite active efforts, there is a gap between committed and achieved targets to meet the goal of 1.5 °C. A study conducted by the IPCC suggests a reduction in global emissions by 45 % (of 2010 emissions) on or before 2050 to achieve this ambitious target. The Paris Agreement has provided a concrete start and first positive step to mitigate the climate crisis.

Kyoto Protocol

The Kyoto Protocol was the first international treaty to legally bind the developed countries to reduce their emissions. Post-industrialisation developed countries have emitted a lot of GHGs into the environment. So, considering this aspect from an equality perspective, it categorised nations into two categories, viz, developed countries and developing countries. The developed countries were legally bound to reduce their emissions by 5.2% below 1990 levels (EU 8%, US 7%, Japan 6%, averaging 5.9%). The Kyoto Protocol also introduced many mechanisms like the Clean Development Mechanism, Emission Trading, and Joint Implementation. Despite its binding obligations, major emitters like the US and other developed countries did not comply with it; on the other hand, being developing nations, China and India were not assigned the binding targets.

UNFCCC

The United Nations Framework Convention on Climate Change (UNFCCC) was established in 1992 at the Rio Earth Summit to reduce greenhouse gas emissions and stabilize the climate system. It was ratified by 198 countries, ensuring the principle of equality and differentiated responsibilities. It was agreed that climate change is a global issue, but countries with different historical backgrounds should bear the responsibilities accordingly. This treaty laid the basis for the Kyoto Protocol and the Paris Agreement. Under its umbrella, the annual Conference of the Parties (COP) is organised to discuss the implementation of previous targets and decide the new targets. This platform has played a pivotal role in shaping the world's climate policies across the globe over the last three decades.

IPCC

The Intergovernmental Panel on Climate Change (IPCC) was established in 1988 by the UNEP and the World Meteorological Organization (WMO). It analyses and reviews thousands of studies and provides comprehensive reports to better understand the climate and climate change and its impacts, along with the mitigation strategies. IPCC has helped to shape many international policies, including the Fourth Assessment Report (2007) and the most famous Special Report on 1.5 °C (2018). A scientific consensus is provided by the work of the IPCC.

Together, the Paris Agreement, the Kyoto Protocol, UNFCCC, and IPCC are the skeleton of the international climate framework and regulations. They have encouraged and mandated the nations to reduce their GHG emissions, along with the flexibility of carbon trading to help developing nations prosper.

Other Regional Treaties

The international treaties, like the Kyoto Protocol and the Paris Agreement, set global targets for climate action. Regional treaties help translate these treaties to suit geographic, economic, and cultural contexts. These regional treaties help member countries to tackle similar local challenges, share knowledge, and combat local vulnerabilities. One of the most famous regional treaties is the European Union's Green Deal. It was launched in 2019 with an ambition of making Europe climate-neutral by 2050. It has played a significant role in shaping the national policies of member countries in the field of the renewable energy sector, biodiversity protection, sustainable industries, and the circular economy. It has proposed binding climate targets to reduce the GHG emissions by 2030 to at least 55% of 1990 levels.

ASEAN Climate Change Initiative (ACCI) is an Asian organisation of ASEAN (Association of Southeast Asian Countries) that is a voluntary platform under ASEAN, promoting steps for climate resilience and sustainable development while focusing on renewable energy, energy efficiency, and disaster management.

African Renewable Energy Initiative (AREI) is the regional collaboration of African countries with the aim of harnessing vast renewable energy, universal energy access, reduction of emissions, and socio-economic development of the region. Cross-border infrastructure projects have also been initiated for climate actions, along with development activities.

The Pacific Alliance and Union of South American Nations [UNASUR *(It is a short form of the Spanish name: Unión de Naciones SURamericanas)*] are the key organisations to promote climate action in the Latin American region. Since 2018, UNASUR has been inactive, but the Pacific Alliance and Amazon Cooperation Treaty Organization (ACTO) are active in this area. Since Amazon's forests are important in carbon sequestration, these countries emphasise forest restoration and forest conservation through these organisations. Nordic countries like Sweden and Norway have successfully implemented carbon pricing and emission trading systems. Regional banks such as the Asian Development Bank (ADB) and the African Development Bank (AfDB) are the major financiers for the regional green infrastructure.

> ***Thinking Corner**: Treaty or Trickery?*
>
> *Do you remember the Ozone hole crisis? The Montreal Protocol succeeded in phasing out the ozone-depleting gases. Why can't we replicate the same for carbon emissions?*

12.3 The Role of Government and Institutions

Governments and organisations are important stakeholders in fulfilling the goals of Green Technology adoption. They can stimulate sustainable growth through policy framing, innovations, investment, and widespread education. Proactive support and steps from governments and institutions can fuel the speed and scale to transition towards the green economy. Let us explore it in detail.

- Policy Framework: The Government and institutions draft policies for the implementation of the green economy. These frameworks and policies are the backbone to direct the national and/or regional efforts to achieve the targets for the green economy and sustainable development. NGOs and other institutions encourage people and societies, while governments set energy efficiency standards, emissions targets, and renewable energy production targets.
- Tax incentives: The Government may introduce direct tax credits, subsidies, and public spending to deploy green technologies. For example, people may be motivated to purchase electric vehicles by subsidizing EVs. In the same way, people may be encouraged to install rooftop solar power plants to sell the electricity to DISCOM companies. Many countries are also providing investment tax credits or production tax credits so that the selling price of solar PV panels may be reduced, and common people may afford them. Thus, contributing to the generation of green energy.
- India's National Solar Mission: Many countries have understood the call of the Paris Agreement and initiated the actions. India's National Solar Mission is one such noteworthy example. In 2010, it was launched as a part of the National Action Plan on Climate Change (NAPCC) to reduce the dependence on fossil fuels. Tariffs, guarantees, subsidies, and government motivations have helped in the development of domestic manufacturing and solar parks. Due to India's national solar mission, India has emerged as a global leader in the solar market worldwide.

12.3.1 The Role of Businesses and Industries

A global journey towards sustainability will not be possible without the active involvement and participation of the industrial sector, as they are the major energy consumers, resource consumers, greenhouse gas emitters, and waste producers. Being an economic powerhouse and the skeleton of development, they are directly affecting resource use, public behaviour, emissions, and innovations. Through Corporate Social Responsibility (CSR), industries participate in promoting social services. It may be in the form of allocating benches, fans, or water coolers to schools, old-age homes, and hospitals. It may be planting trees, donating services, materials, or seminars. Many industries also focus on Environmental, Social, and Governance (ESG) programs to address the sustainability criteria to reduce their carbon footprint. The private sector also boosts innovations and R&D in the field of Green Technology to reduce environmental impact. It may be in the form of biodegradable packaging material, higher energy density rechargeable batteries, smart grids, and electric vehicles.

Green technologies are transforming the existing traditional industries into more efficient, sustainable, and responsible, eco-friendly industries. For example, advancements in battery and power electronics technology have revolutionized public transportation through electric vehicles. Vehicle-to-Grid (V2G: A concept in which Vehicle batteries are used to deliver power to the grid during peak hours), integration of renewable energy sources, and zero emissions are the key advantages of this technology. On the other hand, the Fast-Moving Consumer Goods (FMCG) industry is in the transition stage to replace the packaging materials and product design.

12.3.2 Role of the Common Public:

Sustainable development is not the responsibility of only governments, industries, corporate houses, and NGO type organisations, but common people will also have to participate in this movement of paradigm shifting. Being a common public, we shall also have to change our habits, behaviour, working style, communication, and mentality, because it is a combined responsibility to save our planet, ecology, and environment for our future generations. Grassroots movements, community empowerment, and awareness are required at the level of society and the street to ensure the participation of common people. Steps taken by the average person may be small, but when they attain a mass practice, the results may be drastic. The following are a few steps that common people can contribute to sustainable development:

1. Avoid using artificial lights during the day. Use light coloured paint for walls, use direct lighting, and use energy-efficient lighting sources like LEDs.
2. Use BEE-certified (or other applicable standard) appliances to conserve electricity.
3. Switch off the lights, fans, AC, TV, computers, and other appliances when not in use. Use only those lights or fans where you are sitting in a big hall.
4. Set the temperature of the AC above 25 **°C**, so that it may just condition the air, while saving electricity. Cooling optimization drastically reduces the power consumption.
5. Use proper insulation material to conserve the heat inside the building (in colder regions).
6. Plan a house or a building such that it can utilize natural ventilation to the maximum level. Keep a sufficient number of wide windows, doors, and exhaust holes for the escape of hot air.
7. Plan a rooftop solar power plant to generate electricity for your own needs.
8. Avoid washing machines, water pumps, cereal grinders, and other appliances during the peak-load time.
9. Rather than using the dryer, insist on drying the clothes naturally.
10. Maintain all electrical appliances.
11. Reduce phantom load (energy wasted by devices in standby mode) by unplugging or switching off the power for standby devices.
12. Save water by fixing leaky taps and using water-saving fixtures. Use a bucket instead of a hose for washing cars.
13. Install rainwater harvesting to save rainwater for a year.
14. Avoid single-use plastics. Use cloth bags for vegetables and groceries. Insist on demanding non-plastic containers and packages during shopping.
15. Rather than throwing in garbage, insist on recycling of papers, plastics, and metal waste.

16. Promote safe disposal of e-wastes like batteries, computers, mobile phones, tablets, laptops, TVs, ACs, and other hazardous equipment where toxic minerals have been used during manufacturing.
17. Donate or repurpose old clothes and other non-usable items.
18. Always insist on a public transportation system, like buses and trains.
19. Prefer an electric scooter or an electric car rather than buying a traditional petrol, diesel, or CNG vehicle.
20. Eat seasonal and local foods and fruits.
21. Increase the vegetable part in the food dish. Reduce meat.
22. Plant trees at home, office, terrace, and public places to increase cooling and conserve the environment.
23. Reduce carbon footprint by limiting air travel, buying products with minimal packaging, and supporting companies that follow and promote sustainability.
24. Educate family members and friends about environmental issues.
25. Use biodegradable products, avoid harmful cosmetics, and products that have harmful chemicals.
26. Insist on using eco-friendly construction materials.
27. Control on the purchasing of an excessive number of clothes and other rarely used household items.
28. If you are a farmer, avoid using pesticides. If you are a milkman, sell pure milk. Be a good human for others, too!
29. The list is endless. The moral of the story is that we shall have to practice sustainability and Green Technology in our day-to-day lives. It is ultimately the mass of people whose efforts contribute drop by drop to form a sea!

At this instance, we may recall the Chipko Movement (Hug the Tree movement) of India. It was started in 1970 to conserve the forests. It involved the local villagers, who used to hug the trees to protect them from logging. The Green Belt Movement of Kenya was inspired by Wangari Maathai, a Nobel laureate, who inspired women to plant millions of trees. In Brazil, Amazon forests are conserved by local Indigenous people, helping to protect the ecosystem and cultural heritage. The culture of sustainability can be fostered by awareness programs, education, and mass mobilization of social media to train people for the conservation of our environment and resources. Campaigning can also boost awareness. For example, World Environment Day by the United Nations or Earth Day engages millions of people across the globe to create a sense of shared Responsibility. Green Technology and sustainable development should also be a part of the school and college curriculum to increase awareness.

12.4 Challenges for the Adoption of Green Technology

We have been discussing the various aspects of Green Technology and sustainable development for the last ten chapters. Till now, it has been clear that Green Technology and sustainable practices are essential for human survival on this planet, but there are many challenges in the implementation of Green Technology practices. Now, in this section, let us briefly discuss these challenges and barriers. One of the most prominent barriers is the high initial cost of green technologies in comparison to traditional technologies, particularly for small businesses. However, government subsidies and tax incentives motivate people to go for green technologies. Long-term low operational costs and profit also sometimes attract people to install solar power plants, wind power plants, and others. Access to affordable financing is also a major barrier to the implementation of green technologies, particularly in low-income countries. Underdeveloped financial markets, high interest rates, weather risks, insurance costs, and fear of implementing recent technologies are also barriers. Lack of a clear framework and policies from the government, rigid administration, and bureaucratic attitude are also barriers to Green Technology. At present, very few options are available for green bonds, and concessional loans and risk guarantees also constrain the development of Green Technology. Implementation of Green Technology requires a skilled workforce for installation, operation, and maintenance. Many countries lack such a skilled workforce.

12.5 Emerging Opportunities in Green Technology

The field of Green Technology is rapidly growing to reduce emissions, optimize resources, and promote sustainable development. Green Technology is driven by AI, big data, ML, and blockchain, which are opening new horizons of unseen opportunities in this field, including economic growth and job opportunities.

1. Advancements in biotechnology enable more efficient agriculture and waste management.
2. Innovations in genetic engineering and microbial engineering help provide chemical-free foods and crop resistance to pests.
3. Cleaning of contaminated soil and water with microbes.
4. Development of biofuels from algae and other microorganisms and plants to become an alternative to fossil fuels. At present, many countries, including India, have already started to add 20% plant-based ethanol to petrol.
5. Lots of jobs are expected in the emerging areas of clean energy, waste management, recycling, electric mobility, and sustainable agriculture sectors.
6. The carbon capture and utilization sector promises vast options in carbon-to-products and carbon sequestration.
7. Hydrogen and fuel-cell technology are promising alternatives in the transportation sector.
8. Energy storage solutions like solid-state batteries, green hydrogen storage, and second-life EV batteries are also inspiring many investors for bright opportunities.

9. Floating solar power plants, improved designs of wind turbines, tidal, and wave turbines are at the forefront of the renewable energy sector.

Activity:
Let us calculate the savings on the electricity bill by changing the habit.
Step 1. Note down the reading of your electricity meter at 07.00 pm.
Step 2. Follow your normal routine of using lights and ACs.
Step 3. Note down the meter reading at 07.00 am the next morning. Calculate the difference to get the energy consumption for one night.
Step 4. Now again, note down the reading of your electricity meter at 07.00 pm, like yesterday.
Step 5. Use only necessary lights, switch off the fans, and lamps that are not required. Keep the AC temperature at, say, 27°C, along with a medium speed fan.
Step 6. Note down the meter reading at 07.00 am the next morning. Calculate the difference to get the energy consumption for one night like yesterday.
Step 7. Observe the reduction in electricity consumption for two different nights.
Step 8. Multiply it by 365 to get annual savings in electrical units.
Step 9. Multiply the answer of step 8 by 7 to get annual savings in rupees. (Assuming Rs 7 per unit for the electricity selling rate)
Step 10. Multiply the answer of step 8 by 0.7 to find the reduction in CO_2 *by your efforts.* *(0.7 kg* CO_2 *per unit is India's average grid emission factor, source: CEA Baseline database, version 19,2023)*

Do you know, if you have 3 ACs, you can save 5500 units, Rs ~3,800, 2 tons of CO_2 *equivalent to 98 mature trees!*

Hurray, you did well!

-: EXERCISE:-

Section A: Short Answer Questions

1. List important outcomes of the Paris Agreement.
2. Mention any three UN SDGs related to the environment.
3. Define climate finance.
4. Discuss the role of the IPCC in framing global climate policy.
5. Name any two international organisations that promote Green Technology.
6. Discuss "Common but Differentiated Responsibilities" (CBDR),
7. List outcomes of COP26.
8. Define carbon trading.
9. Discuss the role of the International Solar Alliance (ISA).
10. List any two successful global Green Technology initiatives.
11. List a few regional treaties for climate actions.
12. Discuss the impact of the Kyoto Protocol.

Section B: Long Answer Questions

1. Explain the importance of the Kyoto Protocol and the Paris Agreement.
2. List the achievements of the Paris Agreement.
3. List major challenges faced by developing countries like India in implementing the global sustainability policies. Suggest suitable remedies for the same.
4. List the objectives of the International Solar Alliance and its relevance to India.
5. Discuss the role of international financial mechanisms in combating climate change.
6. Discuss the integration of the United Nations' Sustainability Goals in the protection of environmental and social equity.
7. Explain the global north-south divide in responsibilities and resources for climate action.
8. Explain in detail the significant contributions of India in global climate issues and its alignment of national policies towards mitigating climate change.
9. Elaborate on the role of tax incentives to promote green technologies.
10. Suggest a suitable plan to promote innovation, technology transfer, and equity in climate action for inclusive development.

11. List the emerging opportunities in Green Technology. Propose a suitable career plan for an aspirant in the international climate action department.
12. List 10 common changes in your life that can decrease your carbon footprint.

Section C: Multiple Choice Q and A

1. The primary goal of the Paris Agreement is:
   a) Elimination of all greenhouse gas emissions by 2030
   b) Limit global temperature rises to below 1.5°C above pre-industrial levels
   c) Promote deforestation for economic growth
   d) Ban fossil fuel usage completely
2. The....... treaty was the first to legally bind developed countries to reduce emissions.
   a) UNFCCC b) Kyoto Protocol
   c) European Green d) Paris Agreement Deal
3. The purpose of Nationally Determined Contributions (NDCs) is:
   a) To increase industrial emissions
   b) To set voluntary national emission reduction targets
   c) To ban renewable energy projects
   d) To promote fossil fuel subsidies.
4. .... mechanism in the Paris Agreement ensures countries increase their climate ambitions over time.
   a) Climate targets b) Ratchet Mechanism
   c) Green funds d) Clean goals
5. The scientific assessments on climate change are provided by:
   a) UNEP b) IPCC c) ASEAN d) ADB
6. .... was the key weakness of the Kyoto Protocol.
   a) It excluded major emitters like the US, China, and India from binding targets
   b) It banned all industrial activities
   c) It had no enforcement mechanisms
   d) It only focused on developing countries

7. ……. regional initiative aims to make Europe climate-neutral by 2050.
   a) ASEAN Climate Change Initiative
   b) European Green Deal
   c) Pacific Alliance
   d) African Renewable Energy Initiative
8. The Green Climate Fund (GCF) supports:
   a) Industrial pollution in developed nations
   b) Financial aid for climate adaptation and mitigation in developing countries
   c) Population control
   d) Fossil fuel expansion in developing nations
9. …… is NOT a challenge in adopting Green Technology.
   a) High initial costs
   b) Lack of skilled workforce
   c) Underdeveloped financial markets
   d) Excessive government subsidies
10. The key role of businesses in sustainability is:
   a) Increasing carbon emissions
   b) Implementing ESG (Environmental, Social, Governance) programs
   c) Opposing renewable energy adoption
   d) Avoiding all technological innovations
11. ……..movement involved villagers hugging trees to prevent deforestation.
   a) Earth Day Campaign b) Chipko Movement
   c) Green Belt Movement d) Love the Tree
12. The African Renewable Energy Initiative (AREI) focuses mainly on:
   a) Expanding coal power plants
   b) Promoting universal energy access through renewables
   c) Supporting oil drilling
   d) Banning all energy projects
13. The public can contribute to sustainability by:
   a) Using single-use plastics excessively
   b) Installing rooftop solar panels
   c) Avoiding public transport
   d) Keeping electronic devices in standby mode always

14. Under the Paris Agreement, the Enhanced Transparency Framework (ETF) ensures:
    a) Elimination of all climate policies
    b) Annual reporting of greenhouse gas emissions
    c) Secret negotiations between countries
    d) Only developed countries report emissions
15. …….. sector is expected to create many jobs in Green Technology.
    a) Smart TVs
    b) Disposable Plastics
    c) Electric mobility and waste management
    d) Green Space technology
16. A major advantage of electric vehicles (EVs) is:
    a) Higher life
    b) Cheaper batteries
    c) Zero tailpipe emissions,
    d) Limited battery life with no recycling options
17. …….. Is it a principle of UNFCCC?
    a) Equal pollution for all members irrespective of past manufacturing
    b) Common but differentiated responsibilities
    c) Developing nations must bear all climate costs
    d) Only high-income countries should act on climate change
18. ……. a key feature of India's National Solar Mission.
    a) Promoting solar vehicles
    b) Reducing dependence on fossil fuels through solar energy
    c) Banning all traditional petrol vehicles to reduce pollution
    d) To promote solar planes
19. ……… Technology is used to capture and store carbon emissions.
    a) Pressing the smokes
    b) Carbon capture and utilization (CCU)
    c) Storing industrial waste
    d) Depositing in the coal fields

20. A major barrier to Green Technology adoption in developing countries is:
    a) lack of workers
    b) High interest rates and lack of financing
    c) Overqualified workforce
    d) No government policies
21. ……. initiative focuses on forest conservation in Latin America.
    a) ASEAN b) Pacific Alliance and UNASUR*
    c) Kyoto Protocol d) American Green Deal
22. The IPCC primarily provides:
    a) Forest cover guide
    b) Scientific consensus on climate change
    c) Financial loans for industries
    d) Weather forecast
23. ….. is a renewable energy source.
    a) Petrol b) Petroleum
    c) Floating solar power d) Diesel
24. ….. is the purpose of carbon pricing.
    a) To increase industrial production irrespective of pollution
    b) To incentivize emission reductions through economic costs
    c) To promote deforestation
    d) To increase fossil fuel subsidies
25. …… movement was led by Wangari Maathai.
    a) Chipko Movement b) Green Belt Movement
    c) Solar Lantern for all d) Friday for Future
26. …… is a key benefit of rainwater harvesting.
    a) Water filtering
    b) Conserving and reusing rainwater
    c) Water cleaning
    d) Water quality enhancement
27. …….. is NOT an ESG principle.
    a) Environmental protection
    b) Social responsibility
    c) Maximizing short-term profits at any cost
    d) Governance transparency

28. A major opportunity in green hydrogen technology is:
    a) Jobs in a new field
    b) Providing clean fuel for transportation
    c) Ignoring renewable energy
    d) Hydrogen storage business
29. The Clean Development Mechanism (CDM) helps to:
    a) To integrate development with cleaning
    b) To allow developed nations to invest in emission reduction projects in developing countries
    c) To set up a cleaning mechanism to clean all villages and cities of the nation
    d) To promote cleanliness in schools and universities.
30. …… is a sustainable practice for individuals
    a) Avoiding public transport
    b) Setting the AC temperature below 20°C
    c) Recycling paper and plastics
    d) Using plastic bags daily

True/False Questions (T/F)

31. The Paris Agreement aims to limit global warming to below 2°C.
32. China and India were exempted from emission targets in the Kyoto Protocol.
33. UNFCCC was established at the Rio Earth Summit in 1992.
34. The European Green Deal aims to make Europe climate-neutral by 2030.
35. The IPCC provides financial aid to low-income countries.
36. The Green Climate Fund supports fossil fuel projects.
37. The Chipko Movement originated in Brazil.
38. Carbon capture technology increases emissions.
39. ESG stands for Environmental, Social, and Governance.
40. Rainwater harvesting increases water wastage.
41. The Paris Agreement was adopted in ....
42. The .... introduced the Clean Development Mechanism (CDM).
43. IPCC stands for ....
44. NDCs are under the Paris Agreement.

45. The European Green Deal aims for a..... % emission reduction by 2050.
46. The Green Belt Movement was founded by ....
47. .... produce zero tailpipe emissions.
48. Carbon pricing puts a cost on .....
49. Rainwater harvesting helps in .... conservation.
50. ......focuses on Environmental, Social, and Governance principles.

Answer Keys:

1. B, 2B, 3.B, 4. B, 5. B, 6. A, 7. B, 8.B, 9. B, 10. C, 11. B, 12. B, 13. B, 14 B, 15. B, 16.C, 17. B, 18.B, 19 B, 20. B, 21 B, 22. B, 23B, 24 C, 25. B, 26. B, 27. B. 28 B, 29. B, 30. B

31.F, 32.T, 33.T, 34. F, 35. F, 36. F, 37. F, 38. F, 39.T, 40.F

41.2015, 42. Kyoto Protocol, 43. Intergovernmental Panel on Climate Change, 44. Nationally Determined Contributions, 45.55%, 46. Wangari Maathai, 47. EVs, 48. Carbon emissions, 49. Water conservation, 50. ESG.

**Q 21. Option B is correct historically, but ACTO (Amazon Cooperation Treaty Organization is more relevant today*

उद्यमेन हि सिध्यन्ति कार्याणि न मनोरथैः।
न हि सुप्तस्य सिंहस्य प्रविशन्ति मुखे मृगाः।।

(Chanakya Niti)

Success comes through effort, not merely by wish. A deer does not walk into the mouth of a sleeping lion. In the same way, we shall also have to put combined efforts for sustainable change rather than just promises, treaties, and conferences. Let us adopt it in our daily lives.

Chapter 13

FUTURE OF GREEN TECHNOLOGY AND CONCLUDING REMARKS

Learning Outcomes:

After reading this chapter, the learner will be able to:

1. Identify the key sectors for innovations in Sustainable Development.
2. Explore future predictions in the field of nanotechnology, smart cities, zero-emission transport networks, and agriculture to promote sustainability.
3. Analyse the possible challenges in implementing green technologies.
4. Envision the role of national and international leaders, youth, and NGOs to promote sustainability at large for the nation and at an individual level to practice in daily life.

Thinking Corner: From Innovation to Introspection.

"Dear Arjuna!
People have reached the Moon, invented miraculous gadgets, but have they discovered the art of living to live with less? O Parth!
Should the future be about consuming smarter or consuming less? "

*Life-Guru Krishna**

13.1 Introduction

After industrialisation, the rate of depletion of resources and the rate of pollution have grown exponentially. Given this, it has become a global priority to check global environmental degradation. Experts state that by the year 2050, humans must achieve a net-zero emissions target to escape the climatic catastrophes. Over and above the willpower of humans, technology is the key to reshaping the existing processes to reduce environmental footprints. This chapter explores the innovations needed, pathways to achieve net-zero emissions, visionary predications, and discusses a few innovative technologies.

*This dialogue is fictional, created by the author for illustrative purposes.

13.2 Key Sectors for Innovations

To fulfil the ambitious targets of limiting temperature rise below 1.5 °C requires innovations spanning across diverse fields. Key sectors are energy production, energy consumption, material management, carbon management, and transportation.

- Renewable Energy Sector: Rooftop solar power, Solar parks, wind power, mini and major hydro power plants are the backbone of a sustainable energy generation system. Solar power generation is the most popular way to integrate the common people into sustainable energy generation. Major research is required to be carried out on folding solar panels, lightweight panels, and more efficient and cheaper panels. Offshore wind power generation has vast energy potential, which needs to be tapped through innovative floating wind turbines (or floating solar panels). Deep exploration for geothermal energy is also required to tap the vast potential of natural Earth heat. Innovative ideas may transform the global deserts into a global powerhouse by using the scorching heat of desert sand and sandstorms. Just like the solar rooftop system, one more revolution is required in *domesticating* wind power. Solar power is available only during the daytime, while the availability of wind power is around the clock. If an affordable, simple wind turbine is designed, then it will be a promising alternative *(say sibling!)* to PV solar power

Real Inspiration

One Man, One Tree!

Murugan, a bus conductor from Coimbatore, planted more than 3000 trees from his salary. He also inspires the passengers riding in the bus by distributing free saplings.

A dedicated efforts form a single person has transformed a barren road into a green corridor.

Source: The Better India, 2021

plants on the terrace of the average person to contribute to sustainable energy production.

- Energy Storage: The major disadvantage of renewable energy sources is their intermittency in energy production. This demerit can be addressed with the help of energy storage, which stores the excess energy generated during less-demand periods for utilisation during peak periods or the unavailability of renewable energy. Present Lead-Acid and Nickel-based batteries are proven, cheaper, and durable, but very heavy, whereas Lithium-Ion batteries have advantages of lighter weight, higher energy density, long cycle life (up to 10,000 cycles), quick response times, and limited maintenance costs, but they are much costly[47]. They also have challenges of thermal runaway and a fear of heavy resource depletion. These batteries are typically used in grid storage, home solar systems, and EVs. Supercapacitors are also a new dimension of energy storage. They have higher power density, fast charging-discharging capabilities, long life cycle, and wide temperature tolerance, but they are costly as of this date.
- Smart Grids: Reduction of power loss, energy efficiency enhancement, meeting the peak demand efficiently, monitoring the real-time power demand, improved load balancing, and better integration of renewable energy sources are possible through the use of AI and IoT, sensors, and AI algorithms.
- Carbon Capture: Carbon Capture, Utilization, and Storage (CCUS) is essential to capture the emissions at the source itself rather than allowing them to escape into the environment. CCUS technology is required to be enhanced to cover the maximum emitting sites.
- Alternative to Plastics: At present, it is hard to find even a single product in our daily life that does not involve plastic. There is an urgent need to search for an alternative biodegradable material to replace plastic.

13.3 Dream Future Predications

We are witnessing a positive revolution due to advancements in technology, as well as its negative impacts on our environment. The integration of technologies and positive steps from nations across the globe will reshape our economy and biosphere by 2050.

- Zero-Emission Transport Network: Driverless autonomous transport, electric vehicles, electrified public transport, hyperloop systems, and hydrogen-based transportation will drastically reduce pollution and greenhouse gas emissions.
- Smart Cities: Today's cities may transform into tomorrow's smart cities, leveraging the advantages of AI, IoT, and big data to enhance energy efficiency, improve traffic flow, smart waste management, and surplus power generation for each building. If green hydrogen production becomes cost-effective, then each house, office, or building may have its own power generator, and transmission lines may become obsolete.
- Industrial Ecosystems: Novel practices of waste reduction and circular economy concepts may lead to operating industries in a closed-loop system, enhancing sustainability.
- Agriculture: We have seen animals helping in ploughing in the past; in the present, we are seeing tractors and other machinery helping us in farming and other work. In the future, we may see drones, sensors, and AI tools to help farmers in farming, reducing water usage, and fertilizer usage. Lab meat and plant-based meat alternatives may reduce the dependence on livestock, significantly cutting methane emissions.
- Space-Based Solar Power (SBSP): It involves the collection of uninterrupted solar radiation and transmitting it to Earth with the help of microwaves or laser beams. This technology may prove to be a large-scale source of clean energy to enhance grid reliability.
- Biotechnology for Sustainability: Biotechnology is a novel and unprecedented field having great capabilities at the micro level to offer solutions for carbon capturing, pollution remedies, and sustainable production. Genetically engineered plants can enhance the carbon sequestration

process and the soil carbon storage capacity. Certain microorganisms may produce biofuels, bioplastics, and chemicals. They can also reduce the waste into nontoxic natural materials within a brief time (similar to the actions performed by enzymes in our digestive system). Certain microbes can be tailored to break down hazardous substances to clean our soils and water, for example, an oil spill.

- Nanotechnology: Nano catalysts can enhance the efficiency and speed of a chemical reaction. Nanomaterials are capable of removing microscopic pollutants, heavy metals, and pathogens from water, ensuring clean water. Nanomaterials can also offer lightweight materials to be used in aircraft to reduce emissions and improve efficiency. Nanostructured materials can elevate the function of batteries, supercapacitors, and solar cells.
- There are endless possibilities in robotics, AI, pharma, and other sectors that can transform our lifestyle and sustainability.

13.4 Fusion Energy: A Myth or Reality?

Fusion and fission reactions are the two types of nuclear reactions. In case of nuclear fission, a bigger nucleus breaks down into smaller nuclei, while in a fusion reaction, two lighter nuclei combine to form a bigger nucleus. In both cases, a tremendous amount of energy is released. The Sun and other stars are getting their energy through the fusion reaction. It is the ultimate and most profound source of energy in the universe, having lots of advantages but with few challenges. The primary fuel for the fusion reaction is the isotopes of hydrogen, the most abundant element in our universe. Deuterium, the isotope of hydrogen, can be extracted from seawater, while tritium, another isotope of hydrogen, can be easily bred from lithium. The main advantages of fusion reaction include massive energy density and no emission of greenhouse gases. It also does not pose the risks of accidents like a fission reaction, which can explode as a bomb and contaminate the peripheral area with radioactive waste, making it unsuitable for hundreds of years.

Despite lots of merits, harnessing fusion reaction is yet a dream. Despite decades of research, researchers have not been able to find a commercial way to generate a net positive energy. The main challenge is to heat the plasma above 150 million degrees Celsius and to confine the hot plasma to sustain the reactions. Many nations are researching this ultimate source of energy. Many technologies are being tested, like magnetic confinement using tokamaks and inertial confinement by lasers. The ITER, an international project of thirty-five countries, is building the world's largest tokamak.

In recent years, a few breakthroughs have been achieved in this direction. In December 2022, "Ignition Experiment" was successfully conducted by the US National Ignition Facility to achieve net positive energy from fusion reaction for the first time. Now, efficient magnetic confinement systems are possible due to advancements in high-temperature superconductors, enabling smaller, cheaper fusion reactors.

Even when fusion energy is considered to be the energy, as of this date, it faces many technological hurdles to reach commercial fusion plants. However, the momentum is accelerating, with the expectation of the commercial production of electricity using fusion reaction by 2050. If realized, it can revolutionize the complete energy system. It will be the biggest breakthrough in human history. Let us hope for the best to achieve this dream shortly!

Latest Fact (May 22, 2025)*: Did you know? Germany's Wendelstein 7X, a magnetically confined fusion reactor, achieved a 43-second plasma run for the first time, setting a new operation record! Peak plasma temperature was 30 million °C. Which material scientist might have used to withstand such a high temperature? Hint: The Answer lies in the question*

Impacts of Emerging Trends:

The fusion energy, biodegradable material, and smart grids integrating renewable energy sources are mutually reinforcing and interconnected. Fusion can provide abundant clean energy to power eco-friendly manufacturing. Biodegradable materials and sustainable production practices can reduce carbon footprint and environmental degradation. Smart grids, hydrogen energy, and renewable energy sources can reduce the impacts of climate change. However, rapid technological adoption may pose certain new risks. Bioenergy plantations may threaten our food security in the future. To reduce the carbon footprint, the government may give huge subsidies to farmers for bioenergy crops, so farmers may not prefer food crops. In the era of AI and IoT, too much automation may create a scarcity of jobs, leading to social imbalance and unrest in society. Rapid transition may also find it difficult for the shift workers in the traditional fossil fuel-dependent industries. The data is the key to the future, but smart grids, digital monitoring, and AI/IoT-based devices threaten data security. Technology needs to be accessible to the marginalised population to ensure inclusive growth. Transparent frameworks are essential to manage these risks.

13.5 Global Collaboration

Sustainability is the global challenge for human existence on this planet. It is an urgent issue to be treated beyond national borders, sectors, and cultures for effective collaboration between nations, civil societies, NGOs, and industries across the world. Our next discussion explores the need and effect of advocacy, global leaders in this field, and the role of education and campaigns. Advocacy and campaigns raise awareness about the issue, mobilize public support, and influence policy. Many NGOs have raised their voice against the mass destruction of the environment through industrialization or mining and have succeeded in shifting public opinion to pressure the governments and industries to adopt stronger sustainability commitments. Greenpeace and the World Wildlife Fund (WWF) are leading NGOs working at an international level.

13.6 Prominent Leaders:

- **Mahatma Gandhiji** had said, *"One person can make a difference, and that's the leader."* A leader can ignite the passion and enthusiasm in common people. He is well known for his contribution to the freedom fight of India, but he was also a strict follower of sustainable practices and lived in alignment with nature. Leaders, scientists, policy makers, and activists play a key role in saving the environment as a mass movement.

- **Shri Narendra Modi ji, Hon'ble Prime Minister of India,** has taken many initiatives like authoring a book, *Convenient Action*, and creating a Climate Change Department for the first time in India. He had advocated "common but differentiated responsibilities" for developing nations. Under his leadership, he is committed to climate targets under the Nationally Determined Contributions (NDCs). PM Modi, along with the French President, launched the International Solar Alliance in 2015 to mobilize solar investments of over $1 trillion by 2030. More than 130 countries are part of it. Under the National Electric Mobility Mission Plan (NEMMP), EVs were promoted through subsidies. Under the Ujjwala Yojana (Clean Cooking Fuel Scheme), more than 90 million LPG connections have been provided to rural women to reduce indoor air pollution and deforestation. Under the PM-KUSUM scheme, solar-powered irrigation pumps are provided to Indian farmers. Under the leadership of Prime Minister Narendra Modiji, India has set an ambitious target of 500 GW non-fossil fuel energy installed capacity by the end of this decade. Giga Solar parks, Mega wind farms, and Green Hydrogen production have been prioritized. As of 2025, India is the fourth-largest renewable energy producer in the world [19]. He has also promoted Green Urban development through the Smart Cities Mission, encompassing the values of sustainable infrastructure and energy-efficient buildings. The Namami Gange Programme has been launched to clean and rejuvenate the holy Ganga River to conserve biodiversity through sewage treatment, solid waste management, and riverfronts. Climate finance, Sovereign Green Bonds, and green budget have been introduced in the 2022 budget to mobilize public and private investments. The National Green Hydrogen Mission was launched in 2023 to achieve a target of five million metric tonnes of green hydrogen annually by 2030 to decarbonize the steel, fertilizer, and transport sectors [12]. In a nutshell, we may say that under Prime Minister Narendra Modiji's leadership, India has shifted its stand from being a reluctant participant to a proactive global leader in sustainability.

- **Shri Nitin Gadkari ji**, Hon'ble Union Minister for Road Transport and Highways, has become a face of transformation for India's efforts in sustainable development and climate action in the transport and infrastructure sectors. His efforts and inclination towards green technologies are significantly contributing to India's climate goals under the Modi government. He has played a vital role in green highways through plantation, beautification, over 1,200 km of eco-friendly green corridors, and promoted the use of waste plastic, fly ash, and industrial wastes in road construction. He has also promoted cold mix technology for rural road construction under Pradhan Mantri Gram Sadak Yojana (PMGSY). He has actively promoted Electric Vehicles, pushed EV adoption in government fleets and public transportation, supported battery swapping, and charging infrastructure to popularize EVs. He has aggressively advocated for biofuel production from sugarcane, rice, maize, and bamboo. He has also supported ethanol distillation plants in sugar mills. A 20 % ethanol blending in petrol is permitted to reduce the dependency on fossil fuels [12]. He has sparked the debate and initiated an interest in the Green Hydrogen Transportation system in India by personally using the Hydrogen Car. He has also been credited for introducing electric ferries and inland waterways to reduce road congestion and diesel use. Under his leadership, Bharat Stage (BS)-VI (Indian equivalent to Euro 6) and vehicle scrappage policies have been implemented to reduce pollution. How can we forget the FASTag facility? He has implemented an auto-toll deduction facility, which has reduced idle stoppage time and conserved fuel while paying the toll tax. Shri Nitin Gadkari ji is widely recognised as a result-driven leader in India's green transition.

- **Dr. Vandana Shiva** is a globally renowned environmental scholar who has significantly contributed to the fields of biodiversity, sustainable agriculture, and climate justice. She had founded the Navdanya Movement in 1987 to promote organic farming and seed rights for farmers. She has started the Seed Bank to preserve the indigenous seeds threatened by GMOs. She has helped more than 7,50,000 farmers to adopt sustainable farming without using chemical fertilizers and pesticides. She has also raised her voice for climate justice, the local economy, and regenerative farming to check industrial farming [48].
- **Dr. Chetan Solanki** is widely known as the Solar Man of India or Solar Gandhi, graduated from IIT Bombay, and served as a professor there. He has founded the Energy Swaraj Foundation to promote local clean energy aligned with Gandhian ideas. He has reached more than 2,50,000 people and delivered lectures in schools and universities to promote green energy. His works aim to energise the communities and bring self-reliance to accelerate India's shift to solar power.

- **Greta Thunberg:** She is a Swedish environmental activist who began her climate-related activities at the age of just 15 years. Her speeches and activism influence political and social issues at the international level. Her Fridays for Future (FFF) movement has inspired millions of students to demand stronger climate action. She has become the symbol of youth climate activism to shape the global climate conversation.
- **Christina Figueres**: She was a former executive secretary at UNFCCC. She was the chief architect of the Paris Agreement. She has promoted inclusive development revolving around the climate goals.
- **Wangari Maathai** (Kenya) was a Nobel laureate, an environmentalist, and the founder of the Green Belt Movement, empowering African women to plant more than 51 million trees, linking reforestation with human rights [49].
- **Vanessa Nakate** (Uganda) was a founder of Rise Up Movement who highlighted African climate challenges and prominently worked on clean energy, education, and women's rights.
- **Teresa Ribera** (Spain) is a minister for ecological transition who played a pivotal role in Spain's coal exit and expansion of renewable energy, advocated for a green social contract within the EU, and for climate justice.
- **David Attenborough** (United Kingdom) is a naturalist who promoted awareness of biodiversity loss and climate change through documentaries like *Blue Planet* and *Our Planet.*
- **Al Gore** (USA) is a former Vice President of the USA, an Environmental advocate who promoted climate action through *The Inconvenient Truth* and mobilized public opinion.

- **Elon Musk, Yvon Chouinard, Antonio Guterres, Luisa Neubauer, Licypriya Kangujam**, and many other people are also working in the field of climate change and sustainability, who are actively contributing to shaping the future of green energy and our ecology.

13.7 A Call to Action

Thinking Corner: Want to become a hero like those mentioned above?

Think of a novel idea, prepare a prototype, test it, and patent it. Get help from the nearest innovation centre if required. Visit the websites of the National Innovation Foundation (nif.org.in), the Indian Renewable Energy Development Agency (ireda.in), or the Atal Innovation Mission (aim.gov.in) for more.

We are facing the extreme effects of climate change, including catastrophic cyclones, damaging deluges, frequent storms, droughts, and intense record-breaking heatwaves. Every year, global warming creates a new record. These escalating events will make the existence of all species, including humans on Earth, impossible if proper actions are not taken. The race to rush for blind development to fulfill the greed of humans will devastate this beautiful planet of the universe. Urgent actions from all stakeholders, including government administration, industrialists, individuals, NGOs, and educational institutions, will play an active role in retaining our existence.

- **Governments** must frame strict policies and regulations, incentivise renewable energy, and promote the circular economy. So-called development should never be given priority over the environment.
- **Industries and Corporations** must accept the responsibility for damaging the environment. They should realize that business is not above human existence. They should incorporate sustainability in their operations and

manufacturing practices, along with innovative practices to reduce waste.

- **Academia and Research Organisations** should translate knowledge into practice. Spread awareness among their students.
- **Individuals, Civil Societies, and NGO** should engage communities to put the 5R principle of sustainability in their routine.

13.8 Final Thoughts

Today, sustainability is no longer a theoretical concept, but has become a question of our survival and the existence of our future generations. This book tries to throw light on various renewable energies, green technologies, global and national policies, and practices to provide readers a comprehensive understanding of sustainability and its importance in an easy, simple, non-mathematical way to motivate the common student to implement the green practices in his or her own life as a daily routine. Today, we are on the threshold of a delicate ecological balance. Today's actions will decide the future of our next generations.

The crux of the whole book may be explained in a single Sanskrit verse.

"स्वयं जीवेम शान्त्या सह, परानपि जीवितुमनुज्ञामः।"

Meaning, we are not alone on this planet; this planet is not just for us. Let us live happily, and let others too live happily"!

May this book spark innovations, actions, and awareness for sustainability to realize the future that honours both people and planet.

-: EXERCISE:-

Section A: Short Answer Questions

1. Define the term 'Future-Ready' Green Technology.
2. List emerging trends in green innovation.
3. Define circular economy.
4. Discuss how AI and nanotechnology can help in a sustainable future.
5. Mention a few important sectors for innovation in green technologies.
6. Define sustainable consumption.
7. Discuss the role of youth in sustainable development.
8. List probable future green jobs.
9. Discuss the central theme of Green Technology and Sustainable Development.
10. Explain the concept of a *Zero-emission transport network.*

Section B: Long Answer Questions

1. Discuss the role of Governments and individuals in mitigating climate change.
2. Discuss the role of smart grids in mitigating climate change.
3. Identify five global leaders who have promoted green technologies.
4. Explain how technology can help in transforming conventional manufacturing.
5. Explain the role of an ideal human to have a non-impactful life for Mother Earth.
6. Reflect on the relevance of the idea "Have we invented ways to live with less?" in today's world.
7. Analyse the role of green jobs and the skills required for the green economy.
8. Discuss the importance of educational reforms and innovation hubs to support the development of future green leaders.
9. Assess the need for a paradigm shift in our mindset from 'luxury to responsible consumption'.

10. Propose a vision for India's roadmap to achieve a sustainable green future through innovation, policy framework, and public participation.

Section C: Multiple Choice Q and A

1. The target year to achieve global net-zero emissions is
   a) 2030 b) 2040 c) 2050 d) 2060
2. ……. is NOT a key sector for innovation in Green Technology.
   a) Energy production b) Plastic manufacturing
   c) Transportation d) Carbon management
3. The major disadvantage of renewable energy sources is:
   a) Maintenance b) Intermittency
   c) Costly d) Limited availability
4. A higher energy density, longer cycle life, and expense are the features of:
   a) Zinc-Carbon b) Nickel-based
   c) Lithium-Ion d) Lead-Acid
5. Reduction in power loss and efficiency of grids can be improved by:
   a) Transmission lines b) Smart Grids
   c) Distribution lines d) Quantum Computing
6. The CCUS stands for:
   a) Carbon Capture, Utilization, and Storage
   b) Carbon Control and Usage System
   c) Clean Carbon Use for Society
   d) Climate Change and Urban Sustainability
7. A potential alternative to plastic is:
   a) Biodegradable polymers b) Ceramics
   c) Carbon fibre d) Synthetic rubber
8. The primary fuel for nuclear fusion reactions is:
   a) Uranium b) Thorium
   c) Hydrogen isotopes d) Plutonium
9. The "Ignition Experiment" was successfully conducted by… for fusion energy in 2022?
   a) China b) USA c) India d) Russia
10. The main challenge in harnessing fusion energy is:
    a) High cost of fuel

b) Limited fuel availability
c) Heating plasma above 150 million °C
d) Radioactive waste

11. The International Solar Alliance (ISA) was launched by:
a) India and Japan
b) India and France
c) World Solar Bank
d) European Solar Union

12. India provides LPG connections to rural women under:
a) PM-KUSUM
b) Ujjwala Yojana
c) NEMMP
d) Pm LPG scheme

13. The Navadanya Movement for organic farming was founded by:
a) Al Gore
b) Vandana Shiva
c) Wangari Maathai
d) Chetan Solanki

14. The Fridays for Future (FFF) movement aims to:
a) Promoting nuclear energy
b) Demanding stronger climate action
c) Supporting industrial farming
d) Advocating for plastic bans

15. .... technology collects solar power in space and transmits it to Earth:
a) Solar rooftops
b) Space-Based Solar Power (SBSP)
c) Solar Space Protocol
d) Solar Substation

16. The main advantage of fusion energy over fission is:
a) No radioactive waste
b) No greenhouse gas emissions
c) Both a and b
d) None of the above

17. Who, as a minister in India, promoted the use of biofuels and electric vehicles?
a) Dharmendra Pradhan
b) Nitin Gadkari
c) Prakash Javadekar
d) Jayant Chaudhari

18. The purpose of the Namami Gange Programme is:
a) Solar energy promotion along the banks of rivers
b) Cleaning the Ganga River
c) Water canal in the Ganga
d) Afforestation along the banks of all rivers

19. Technology can help in cleaning oil spills using microorganisms.
    a) Robotics b) Biotechnology
    c) Nanotechnology d) AI
20. The main risk of rapid bioenergy crop adoption is:
    a) High water consumption b) Threat to food security
    c) Increased pollution d) Deforestation
21. The documentary raised awareness about biodiversity loss.
    a) The Inconvenient Truth b) Blue Planet
    c) Before the Flood d) An Inconvenient Sequel
22. The main goal of the Green Belt Movement is:
    a) Reducing plastic waste
    b) Planting trees and empowering women
    c) Promoting solar energy
    d) Advocating for electric vehicles
23. ….. policy in India aims to phase out old, polluting vehicles.
    a) BS-VI norms b) Vehicle Scrappage Policy
    c) Fastag d) mParivahan
24. The 5R principle of sustainability stands for:
    a) Reduce, Reuse, Recycle, Recover, Refuse
    b) Restore, Regulate, Reclaim, Replant, Reuse
    c) Read, Research, Report, Reform, Rebuild
    d) None of the above
25. The primary objective of the National Green Hydrogen Mission is:
    a) Promote fossil fuels b) Decarbonize industries
    c) Expand nuclear energy d) Expand nuclear energy
26. Battery and solar cell performance can be enhanced by:
    a) Biotechnology b) Nanotechnology
    c) AI d) None of the above
27. "One person can make a difference, and that's the leader," was said by:
    a) Mahatma Gandhi b) David Attenborough
    c) Narendra Modi d) None of the above
28. The main risk of excessive AI and automation is:
    a) Increased pollution b) Job scarcity
    c) Data security threats d) Higher energy consumption

29. Greta Thunberg initiated:
    a) Greenpeace
    b) Fridays for Future
    c) Extinction Rebellion
    d) Navdanya
30. The main purpose of the PM-KUSUM scheme is:
    a) Solar-powered irrigation pumps
    b) Waste management
    c) Electric vehicle subsidies
    d) LPG distribution
31. The net-zero emissions target must be achieved by 2050. (T/F)
32. Lithium-ion batteries are cheaper than Lead-Acid batteries in initial cost. (T/F)
33. Smart grids use AI and IoT for energy efficiency. (T/F)
34. Fusion energy produces lower radioactive waste than fission. (T/F)
35. 35 countries are involved in the ITER project. (T/F)
36. India is the world's largest renewable energy producer as of 2025. (T/F)
37. The Navdanya Movement was started by Greta Thunberg (T/F)
38. The FASTag system helps reduce fuel wastage at toll booths. (T/F)
39. Biotechnology cannot help in carbon capture. (T/F)
40. Christina Figueres led the Paris Agreement. (T/F)

41. The ________ experiment achieved net positive energy from fusion in 2022.
42. ________ is a Swedish activist known for the Fridays for Future movement.
43. The ________ Movement planted over 51 million trees in Africa.
44. ________ is an Indian scheme promoting solar-powered irrigation.
45. The ________ Programme aims to clean the Ganga River.
46. ________ is the primary fuel for nuclear fusion reactions.
47. ________ is a technology that can remove microscopic pollutants from water.
48. The ________ policy in India promotes ethanol blending in petrol.

49. ________ is an international project building the largest tokamak for fusion research.
50. The ________ principle includes Reduce, Reuse, and Recycle.

Answer keys:

1. C, 2. B, 3. B, 4. C, 5. B, 6. A, 7. A, 8. C, 9. B, 10. C, 11.B, 12. B, 13. B, 14. B, 15.B, 16. C, 17. B, 18. B, 19. B, 20. B, 21. B, 22. B, 23. B, 24. A, 25. B, 26. B, 27. A, 28. B, 29. B, 30. A

True/False

31. T, 32. F, 33. T, 34. F, 35. T, 36. F, 37. F, 38. T, 39. F, 40. T

Fill in the Blanks

41. Ignition, 42. Greta Thunberg, 43. Green Belt, 44. PM-KUSUM, 45. Namami Gange, 46. Hydrogen isotopes, 47. Nanotechnology, 48. Ethanol blending, 49. ITER, 50. 5R

Appendix 1

Global Greenhouse Gas Emissions by Sector (2023 World Data)

Sector	Percentage of global GHG emissions*	Main Sources
Energy (Electricity and Heat)	31%	Fossil fuel combustion in power plants and heating
Industry	24%	Cement, steel. Chemicals, mining, and manufacturing
Agriculture, forestry, and land use (AFOLU)	18%	Methane generation from livestock, deforestation, soil use, and fertilizers
Transport	15%	Road, aviation, shipping, rail
Buildings	6%	Heating, cooling, and lighting
Waste	3%	Landfills, wastewater, waste incineration
Other	3%	Other emissions and burning, oil-gas leaks

Appendix 2

India's Greenhouse Gas Emissions by Sector (2023)

Sector	Percentage of Indian GHG emissions*	Main Sources
Electricity Generation	42%	Mainly coal-based power plants
Agriculture	18%	Methane from livestock, rice cultivation
Industries	16%	Cement, Steel, Chemicals
Transport	14%	Mainly road transport
Buildings	7%	Residential and commercial energy use
Waste and Others	3%	Landfills, wastewater treatment

Source: Intergovernmental Panel on Climate Change (IPCC)

**Percentages are approximated due to overlapping boundaries.*

Appendix 3
Emission by Greenhouse Gas Type (2023 World Data)

Gas	% of Global GHG Emissions	Sources
Carbon Dioxide (CO_2)	~74%	Fossil fuel burning, power plants, and industries.
Methane (CH_4)	~17%	Agriculture (livestock, rice), landfills, and fossil fuel use
Nitrous Oxide (N_2O)	~6%	Fertilizers, agriculture, biomass burning
Fluorinated Gases	~3%	Refrigerants, air conditioners, and electronics

Appendix 4
Top Contributors by Country/Region (2023 Data)

Country	% of Global GHG Emissions	Main Sources
China	~30%	Coal power, industries
United States	~13%	Transportation, building
India	~7%	Coal power, agriculture
EU	~7%	Industries, transportation
Rest of the World	~43%	Mixed sources

Sources: IPCC AR6, 2023; IEA World Energy Outlook, 2023; UN Emissions Gap Report 2024

Appendix 5

Global Energy Access (2023)

Indicator	Value	Remark
People without electricity	~670 million	Concentrated mainly in Saharan Africa and a few parts of Asia
People without clean cooking gas	~2.3 billion	Reliant on biomass, kerosene, and other fossil fuels
Global electricity access rate	91%	It was 83% in 2010, so it improved
Rural vs Urban Energy Access Gap	81% rural 99% Urban	There is a major disparity in low-income and developing countries.

Appendix 6

Renewable Mini-Grids and Off-Grid Systems across the Globe

Deployment Type	Households Reached (estimated)	Main Regions
Solar Home Systems	~50 million	Africa, South Asia
Mini-Grids (Mostly solar)	~47 million	Nigeria, Kenya, Bangladesh, Myanmar
Off-Grid Renewables (all)	>110 million	Key driver for rural electrification

Notes: Even when electrification is rapidly expanding, yet 1 in 11 people still lack access to electricity. Sources: World Bank, SDG7 report (2024); IEA World Energy Outlook (2023); SEforALL (Sustainable Energy for All) data platform

Appendix 7

Research Centres and Academic Institutions:

1. Indian Institute of Technology (IITs)- various campuses
2. National Institute of Wind Energy (NIWE)
3. Indian Institute of Science (IISc), Bengaluru
4. National Renewable Energy Laboratory (NREL)
5. The Energy Research Institute, NITI Aayog, New Delhi

Appendix 8

India's Energy Mix (Installed Capacity in GW as of 2024) [19]

Energy Source	Installed Capacity in GW	% Share in Total Capacity
Coal	210	57%
Renewables (Solar, Wind, Hydro, Biomass)	140	38%
Natural Gas	25	7%
Nuclear	7	2%
Total	382	100%

Appendix 9

Renewable Energy Adoption in India (Cumulative capacity in GW)

Technology	2010	2020	2024	Growth % (2015-2024)
Solar PV	4.7	35	70	1390%
Wind	26	38	45	73%
Small Hydro and Biomass	10	12	14	40%

Sources: Various reports and data from websites of the concerned ministries, IEA, TERI, and CEA(India)

Appendix 10

Global Emissions Reduction Targets

Country	Target Year	GHG Reduction Goal	Remark
EU	2030	55% reduction from 1990 levels	Policy implementation underway
USA	2030	52% reduction from 2005 levels	Initiatives are in progress at the federal and state levels
China	2030/2060	Emission reduction by 2030, but Carbon neutrality by 2060	Invest heavily in nuclear and renewables
India	2030/2070	Emission reduction by 2030, but Carbon neutrality by 2070	Aggressive expansion of the renewable sector
Brazil	2030	43% reduction from 2005 levels	Reforestation concerns are gaining priority

Appendix 11

India's National Climate Commitments (NDC under the Paris Agreement) [12]

Commitment	Target Year	Details	Progress as of 2024
Reduce Emissions	2030	45% reduction from 2005 levels	Achieved ~35 % reductions
Increase non-fossil fuel capacity	2030	500 GW renewable energy capacity	Installed capacity ~450 GW
Create Carbon Sink	2030	2.5 to 3 billion tonnes CO_2 equivalent by forests	Afforestation and reforestation programs are ongoing
Achieve Net Zero carbon	2070	India's long-term goals are announced at COP26	Policy framework being developed

Appendix 12

Progress on Key Energy- Environment Indicators (2023)

Indicator	Global Average	India's Status	Target Year
Population with Electricity Access	91%	>99%	2030
Renewable Energy Share in Total Energy	17%	11%	2030 (to reach 40%)
CO_2 emissions per capita	4.8 t	1.9 t	Significant reduction
Waste Recycling Rate	14%	10%	To further increase

Sources: UN Sustainable Development Goals Report 2024, IEA World Energy Outlook 2024, Ministry of Statistics and Programme Implementation (India). IRENA Renewable Energy Statistics 2024 [4].

Appendix 13
Sustainable Development Goals and Key Indicators

SDG	Goal Description	Environment Indicators
SDG 7 Affordable and Clean Energy	To ensure reliable, affordable, and clean energy for all.	1. Number of households with access to electricity 2. Share of renewable energy in total energy generation [11]. 3. Greenhouse gas emissions per capita
SDG 13 Climate Action	To mitigate climate change	1. Number of nations with climate action plans 2. Carbon intensity of GDP 3. Greenhouse gas emissions per capita
SDG 12 Responsible Consumption and Production	To ensure sustainable production and consumption	1. Material footprint per capita 2. Recycling rate 3. Energy efficiency in industries
SDG 6 Clean Water and Sanitation	To avail the facility of clean water and sanitation	1. Percentage of water used in industries. 2. Access to safe drinking water 3. Access to proper sanitation
SDG 11 Sustainable Cities and Communities	To design safe, energy-efficient, sustainable, and resilient cities.	1. Use of public transport rates 2. Urban forest per capita 3. Air quality index in urban areas

Appendix 14

India's Energy Access and Electrification Status

Indicator	Status as of 2024	Remark
Population with Electricity Access	>99%	Almost near-universal access
Rural Electrification Rate	~98%	Significant improvement from 2010
Availability of clean cooking fuels	~60%	Rural areas still rely on the woods
Households with solar systems	~ 50 lakhs	Rapid expansion in recent years

Appendix 15

Electrification Progress in India (2010-2024)

Year	Total %	Rural Electrification	Urban Electrification
2010	79%	60%	99%
2015	85%	75%	99.5%
2020	97%	93%	99.9%
2024	>99%	98%	100%

Sources: International Energy Agency (IEA) World Energy Outlook 2024, Ministry of Power, Government of India (2024), World bank Energy Access Reports (2023), Sustainable Energy for all (SEforALL) Reports 2024

Appendix 16: List of Suggested Practical

1. To test the efficiency of LED vs traditional light sources like incandescent bulbs, tube lights, and CFLs.
2. To measure the solar irradiance using a pyrometer at various times and months. Compare the same.
3. To design a simulation of a solar photovoltaic system using software for a rooftop solar power plant of a home or an institute.
4. To calculate the carbon footprint of a student and an institute using online tools based on electricity, transportation, and food usage. Calculate the equivalent number of trees to be planted per annum to offset the carbon footprint.
5. To demonstrate the vermicomposting of organic waste, monitoring its temperature and moisture.
6. To create a grey recycling working model to treat grey water for gardening use.
7. To design a rainwater harvesting system. Avail information about the rooftop area and average rainfall per annum. Calculate the size of the underground water tank required.
8. To verify the time, quality of food, and energy efficiency of the traditional pressure cooker and solar cooker system.
9. To analyse your institute building or house to assess the heat/cold leakage. Suggest steps to improve HVAC efficiency.
10. To conduct a field survey in a nearby forest/park/mountain to identify the local birds, plants, animals, trees, and insects. Link your findings to the UN SDGs to appreciate the biodiversity of your local ecosystem.
11. To analyse the local air quality in your campus and city for SPM, PM 2.5, NOx, and Sox. Suggest mitigation.
12. To create a small working model of a biogas plant or plan a visit to a nearby biogas plant. Guess the methane output per kg of waste.
13. To prepare a miniature wind turbine model. Test it for different blade angles, and measure the voltage generation using a multimeter.
14. To test the water, you are drinking for pH, turbidity, and TDS. Compare the results with BIS standards. Suggest suitable measures.
15. To carry out the preliminary energy audit for your institute, identify the inefficiencies, recommend suitable measures.

Appendix 17
Overview of ISO 50001- Energy Management System Standard

ISO 50001 is an international standard developed by the International Organization for Standards (ISO) to establish, implement, maintain, and improve the energy management system (EnMS). It helps the industries and organisations to improve energy performance. It was first published in 2011 and revised in 2018. It follows the cycle of Plan-Do-Check-Act (PDCA) for the overall improvement in energy efficiency and the environment. It also ensures the energy baseline and energy performance indicators (EnPIs) by conducting energy reviews in the potential area of energy use. It sets the energy objectives and action plans to reduce the energy consumption and thus the carbon footprint and greenhouse gas emissions. It helps organizations to realize the untapped potential in energy efficiency.

Key Facts:

- No quantitative targets specified, the organisation fixes the targets and plans steps to achieve the targets.
- Development of energy policy
- Energy planning and regular reviews
- Compliance with regulatory frameworks and legal requirements
- Energy management programs for energy monitoring, measurement, and analysis to achieve continuous improvement
- Certification system to ensure customer, employee, and stakeholders.

Benefits of ISO 50001 implementation:

- Improved energy efficiency for various processes and facilities, Reduction in energy demand. Thus, reduced dependence on fossil fuels.
- Improvement in climate action goals.
- Reduced greenhouse gas emissions and carbon footprint of the organisation.

- Better contribution towards sustainability and UN SDGs 7 and 13.
- Implementation of environmental regulations and government guidelines, thus increasing the chances of appreciations/certificates/awards from climate-related NGOs, national/international organisations, and the government.
- Enhanced reputation as a socially responsible organisation.

ISO 50001 also complements other initiatives like LEED, IGBC, and BREEAM certifications. It can also be integrated with ISO 9001 (for quality management) and ISO 14001 (for environmental management) standards.

Appendix 18

Sample Paper for Practice

University Final Examination for all UG Programs

Subject: Green Technology and Sustainable Development
Time: 3 Hours **Max. Marks: 100**

PART A: Very Short Answer Questions.
Each question carries 1 mark.
All questions are compulsory.
Q. 1 Define Green Technology.
Q. 2 List any two UN Sustainability Goals.
Q. 3. Name the gas responsible for acid rain.
Q. 4. State the full form of LEED.
Q. 5 Name any one carbon capture site in India.
Q. 6 Define 'Triple Bottom Line'.
Q. 7 List two benefits of electric vehicles.
Q. 8 Name the material used for manufacturing PV cells for solar power generation.
Q. 9 State one sustainable farming practice.
Q. 10 State the name of a city that plans to be carbon neutral by 2025.
Part B: Q. No. 11 to 15 Short Answer Questions. Each question carries 2 marks.
Q. 11 Compare the Green economy and the Traditional economy.
Q. 12 State the role of IGBC in India.
Q. 13 Explain with one example, Industrial Symbiosis.
Q.14 Mention any two goals of the Paris Agreement.
Q. 15 Explain in brief: Net-Zero-Emissions.
Part C: Q. No. 16 to 25 Medium Answer Questions. Each question carries 3 marks.
Q. 16 State key features of the circular economy.
Q. 17 Mention the role of wind energy in sustainability.

Q. 17 Compare PV and CSP types of solar technology for electricity generation.

Q. 18 Discuss the positive and negative impacts of the Green Revolution.

Q. 19 List three international collaborations to mitigate climate change.

Q. 20 Explain how AI can be used in improving energy efficiency.

Q. 21 List and explain each component of the wind power system.

Q. 22 Discuss three features of the LEED certification system.

Q. 23 State the advantages of the Smart Grid.

Q. 24 Define Green Manufacturing. State its importance in planetary health.

Q. 25 Explain the role of the government in promoting electric vehicles.

Part D: Q. No. 26 to 35 Long Answer Questions.

Each question carries 5 marks. Attempt ANY FIVE

Q. 26 Discuss any two initiatives by the Government of India to promote green technologies.

Q. 27 Discuss geothermal energy and its importance in mitigating climate change.

Q. 28 Explain the need for energy conservation in ICT.

Q. 29 Discuss the importance of the International Solar Alliance in the present era.

Q. 30 Explain the role of green buildings in improving energy efficiency.

Q. 31 Describe the role of India in promoting renewable energy worldwide.

Q. 32 Explain the challenges of a large hydropower plant.

Q. 32 Discuss how 'fast fashion' is unsustainable.

Q.33 Explain the role of water conservation and water harvesting in environmental sustainability.

Q.34 State several types of hydropower plants. Explain any one in detail.

Q. 35 Discuss the role of green chemistry in improving sustainability.

List of various Abbreviations used in the book.

AC	Alternating Current
AD	Anno Domini
AFOLU	Agriculture, Forestry, and Other Land Use
AI	Artificial Intelligence
AMI	Advanced Metering Infrastructure
AREI	African Renewable Energy Initiative
BAS	Building Automation System
BEE	Bureau of Energy Efficiency
CAISO	California Independent System Operator
CCS	Carbon Capture and Storage
CCUS	Carbon Capture, Utilization, and Storage
CEA	Central Electricity Authority
CH_4	Methane
CIF	Climate Investment Funds
CO_2	Carbon Dioxide
COP	Conference of the Parties
CPCB	Central Pollution Control Board
CSP	Concentrated Solar Power
CSIR	Council of Scientific and Industrial Research
DERs	Distributed Energy Resources
EESL	Energy Efficiency Services Limited
EIA	Environmental Impact Assessment
EnMS	Energy Management System
EPR	Extended Producer Responsibility
ESG	Environmental, Social, and Governance
ESS	Energy Storage System
EU	European Union
EUETS	European Union Emissions Trading System
EV	Electric Vehicle
FiTs	Feed-in Tariffs
GCF	Green Climate Fund
GEF	Global Environment Facility
GHG	Greenhouse Gas
GNH	Gross National Happiness

GRI	Global Reporting Initiative
GSCM	Green Supply Chain Management
GW	Gigawatt
HVAC	Heating, Ventilation, and Air Conditioning
IC	Internal Combustion
IEA	International Energy Agency
IES	Indian Engineering Services
IIT	Indian Institute of Technology
IISc	Indian Institute of Science
INDC	Intended Nationally Determined Contribution
INR	Indian Rupee
IoT	Internet of Things
IPCC	Intergovernmental Panel on Climate Change
IREDA	Indian Renewable Energy Development Agency
IRENA	International Renewable Energy Agency
ISA	International Solar Alliance
ISTE	Indian Society for Technical Education
JET	Joint European Torus
JIT	Just-In-Time
kWh	Kilowatt-hour
LCD	Liquid Crystal Display
LCOE	Levelized Cost of Electricity
LED	Light Emitting Diode
LPG	Liquefied Petroleum Gas
MJ	Megajoule
MNRE	Ministry of New and Renewable Energy
MoEFCC	Ministry of Environment, Forest, and Climate Change
MW	Megawatt
N2O	Nitrous Oxide
NDC	Nationally Determined Contribution
NDRC	National Development and Reform Commission
NEEPCO	North-eastern Electric Power Corporation
NGO	Non-Governmental Organization
NH_3	Ammonia
NIWE	National Institute of Wind Energy
NISE	National Institute of Solar Energy

NITI Aayog	National Institution for Transforming India
NO_x	Nitrogen Oxides
PHS	Pumped Hydro Storage
PM	Prime Minister
PV	Photovoltaic
R&D	Research and Development
REEs	Rare Earth Elements
RPS	Renewable Portfolio Standards
SDG	Sustainable Development Goal
SEforALL	Sustainable Energy for All
SERI	Solar Energy Research Institute
SMR	Small Modular Reactor / Steam Methane Reforming
TBL	Triple Bottom Line
TES	Thermal Energy Storage
TERI	The Energy and Resources Institute
UN	United Nations
UNDP	United Nations Development Programme
UNEP	United Nations Environment Programme
UNFCCC	United Nations Framework Convention on Climate Change
USD	United States Dollar
UV	Ultraviolet
VFD	Variable Frequency Drive
VOC	Volatile Organic Compound
WEEE	Waste Electrical and Electronic Equipment
W/m^2	Watts per square meter
WWF	Worldwide Fund for Nature

Glossary of Technical Terms

Term	Definition
5R Principle	It is a concept of the 5Rs. This includes Refuse, Reduce, Reuse, Repurpose, and Recycle. It is a sustainable way to reduce waste.
Circular Economy	It aims to have a closed system to eliminate waste and maximise resource use through the 5R principle.
Climate Change	An alteration of the weather and temperature due to greenhouse gas emissions.
Concentrated Solar Power (CSP)	It is a technology to concentrate sunlight using mirrors to generate electricity or heat.
Decarbonization	It refers to the process of reducing carbon dioxide in the system.
Green Building	A building designed to use minimum resources.
Greenhouse Gases (GHG)	Those gases that trap heat and increase the global temperature.
Hydrogen Fuel Cell	It is a device to convert hydrogen and oxygen into water while generating electricity. No pollution is associated with this process.
Insolation	It is the amount of solar energy received per unit area at a given time.
Net-Zero Emissions	It refers to the balanced state, where generated and absorbed GHG are equal. So, no new net emissions are released into the atmosphere.
Passive Solar Design	It is a design in which no active additional energy is required to be supplied for heat and light.

Photovoltaic (PV)	It is technology that converts sunlight directly into electricity using semiconductor materials.
Renewable Energy	It refers to those natural energy sources that continuously replenish and do not vanish with time or use.
Sustainable Development	Present development without compromising the ability of future generations.

Key Websites for Green Technology and Sustainable Development

Sr. No.	Agency	Website	Description
1	International Renewable Energy Agency (IRENA)	https://www.irena.org	Detailed data and reports on renewable energy technologies, policies, and global trends
2	International Energy Agency (IEA)	https://www.iea.org	Detailed statistics, outlooks, and a comprehensive roadmap on renewable technologies, energy efficiency, and sustainable development
3	United Nations Environment Programme (UNEP)	https://www.unep.org	Green Technology adaptation, sustainability, and global environmental issues
4	National Renewable Energy Laboratory (NREL)	https://www.nrel.gov	Latest research and development about solar, wind, and other

			renewable energy sources
5	Solar Energy Industries Association (SEIA)	https://www.seia.org	Latest market trends, industrial news, and policy updates on solar energy
6	International Solar Alliance (ISA)	https://isolaralliance.org	Solar projects in tropical countries
7	Global Wind Energy Council (GWEC)	https://gwec.net	Latest information on wind technologies, wind innovations, and wind policies.
8	The Climate Group	https://www.theclimategroup.org	Collaborations and campaigns on sustainable business practices and renewable energy

List of Organisations Promoting Green Technology

International Agencies:

1. United Nations Environment Programme (UNEP)

2. International Renewable Energy Agency (IRENA)

3. International Panel on Climate Change (IPCC)

4. International Energy Agency (IEA)

5. Global Environment Facility (GEF)

6. United Nations Development Programme (UNDP)

7. International Solar Alliance (ISA)

Indian Agencies:

9. Ministry of New and Renewable Energy (MNRE)
10. Central Electricity Authority (CEA)
11. Bureau of Energy Efficiency (BEE)
12. Indian Renewable Energy Development Agency (IREDA)
13. National Institute of Solar Energy (NISE)
14. Council of Scientific and Industrial Research (CSIR)
15. Energy Efficiency Services Limited (EESL)
16. Central Pollution Control Board (CPCB)

Non-Governmental Organizations (NGOs) and Civil Societies (India):

1. The Energy and Resources Institute (TERI)
2. Centre for Science and Environment (CSE)
3. Greenpeace
4. SELCO Foundation
5. Worldwide Fund for Nature (WWF) India

Indian Leaders in Sustainability

Name	Area of Focus	Main Contributions
Shri Narendra Modi, Prime Minister of India	Renewable Energy, Social Inclusion, International Solar Alliance	Strong leadership, to boost India in the renewable energy sector.
Shri Nitin Gadkari, Minister, Road, Transport and Highways, India	Green Energy, Hydrogen, Electric Vehicles	Personal boost to green technologies in the transportation sector
Dr. R. A. Mashelkar	Science and Innovation	Sustainable industrial R&D
Dr. Vandana Shiva	Agriculture	Seed rights, Organic farming
Sunita Narain	Urban environment	Air quality, water, and environmental equity
Dr. Arunabha Ghosh	Policy	Climate finance, SDG research
Suresh Prabhu	Policy and Governance	Sustainability in infrastructure and public utilities
Dr. Ashok Khosla	Community Development	Rural technology, green livelihoods
C. K. Mishra	Climate Diplomacy	Paris Agreement, biodiversity protection
Dr. Chetan Solanki	Climate Awareness	Promotion of solar energy

Global Leaders in Sustainability

Name	Country	Sector	Main Contributions
Christina Figueres	Costa Rica	Diplomacy	Architect of the Paris Agreement
Dr. Sultan Al Jaber	UAE	Policy, Business	COP28, Clean energy diplomacy
Frans Timmermans	Netherlands	Policy	EU Green Deal. Carbon Border Mechanism
Greta Thunberg	Sweden	Atavism	Global youth movement for climate action
Prof. Johan Rockstrom	Sweden	Climate Policy	COP26 Glasgow Climate Pact
Patricia Espinosa	Mexico	Climate Diplomacy	UNFCC leadership
Elon Musk	USA	Technology	Mainstreaming EVs and Battery technology
Dr. Fatih Birol	Turkey	Energy Analysis	IEA's clean energy shift, Net-zero roadmap

References

1. United Nations. Transforming Our World: The 2030 Agenda for Sustainable Development. United Nations, 2015.
2. Wangchuk, Sonam. "Teaching Through Experience." TED Talk, SECMOL Ladakh, 2014.
3. International Energy Agency. World Energy Outlook 2024. IEA, 2024.
4. International Renewable Energy Agency (IRENA). Renewable Energy Statistics 2024. Abu Dhabi, 2024.
5. Bhatia, S.C. Advanced Renewable Energy Systems (Parts 1 and 2). New Delhi: WPI Publishing, 2014.
6. Goel, Malti, Verma, V.S., and Tripathi, Neha Goel. Solar Energy. Springer, 2023.
7. Intergovernmental Panel on Climate Change (IPCC). Climate Change 2023: Synthesis Report. Geneva, Switzerland, 2023.
8. Kumar, Narendra, Singh, Hukum, and Kumar, Amit. Renewable Energy and Green Technologies: Principles and Applications. CRC Press, 2021.
9. United Nations Environment Program (UNEP). Global Environment Outlook 6: Healthy Planet, Healthy People. Nairobi, Kenya, 2019.
10. European Commission. The European Green Deal. Brussels: European Commission, 2019.
11. Sustainable Energy for All (SEforALL). Energy Access and Transition Progress Report. 2023.
12. Ministry of New and Renewable Energy (MNRE). Annual Report 2023-24. Government of India, New Delhi.
13. Chudnovsky, Bella H. Transmission, Distribution and Renewable Energy Generation. Springer, 2022.
14. Singh, R.K., Ojha, C.S.P., and Rai, R.K. Handbook of Applied Hydrology. Springer Nature, 2021.
15. United Nations Framework Convention on Climate Change (UNFCCC). The Paris Agreement. United Nations, 2015.

16. International Solar Alliance (ISA). International Solar Alliance Strategic Plan. New Delhi, 2022.
17. Global Wind Energy Council (GWEC). Global Wind Report 2024. Brussels: GWEC, 2024.
18. Kothari, D.P., Singal, K.C., and Ranjan, R. Renewable Energy Sources and Emerging Technologies. 3rd ed., New Delhi: PHI Learning, 2022.
19. Government of India. India Energy Dashboard 2024. NITI Aayog and Central Electricity Authority, New Delhi.
20. Wikipedia contributors. Green Technology. Wikipedia, The Free Encyclopedia. Available at: https://en.wikipedia.org/wiki/Green_technology [Accessed 15 June 2025].
21. World Commission on Environment and Development. Our Common Future. Oxford University Press, 1987.
22. Fraunhofer Institute for Solar Energy Systems. PV Basics. Available at: https://www.ise.fraunhofer.de
23. International Energy Agency (IEA). Global EV Outlook 2023. Available at: https://www.iea.org/reports/global-ev-outlook-2023
24. Intergovernmental Science-Policy Platform on Biodiversity and Ecosystem Services (IPBES). Global Assessment Report on Biodiversity and Ecosystem Services, 2019.
25. NASA. Climate Change and Global Warming. Available at: https://climate.nasa.gov
26. IPBES. Global Assessment Report on Biodiversity and Ecosystem Services, 2019.
27. Bloomberg NEF. EV Policy Tracker 2024. Available at: https://about.bnef.com/electric-vehicle-outlook/
28. International Federation of Organic Agriculture Movements (IFOAM). Principles of Organic Agriculture. Available at: https://www.ifoam.bio
29. World Health Organization (WHO). "Air Pollution." Fact Sheet, 2023.

Available at: https://www.who.int/news-room/fact-sheets/detail/ambient-(outdoor)-air-quality-and-health

30. United Nations Environment Program (UNEP). "The Disappearing Aral Sea." UNEP Report, 2021. Available at: https://www.unep.org/resources/disappearing-aral-sea
31. Mingyang Smart Energy, *Mingyang Launches World's First 20 MW Offshore Wind Turbine*, Press Release, 2024. Available at: https://www.myse.com.cn/en/news/2024/20MW.html
32. Goldwind, *Goldwind Installs World's Largest Wind Turbine (GWH252-16MW)*, Press Release, 2024. Available at: https://www.goldwind.com/news/2024/16MW-turbine.html
33. Equinor, *Hywind Tampen: The World's Largest Floating Wind Farm*, Company Report, 2025. Available at: https://www.equinor.com/energy/hywind-tampen
34. World Wind Energy Association (WWEA), *World Wind Energy Report 2023*, WWEA, March 2024. Available at: https://wwindea.org/ssuploads/media/2024/3/1711538106-40ab83f2-3e01-4c0a-9d28-e0a21bff72e6.pdf
35. U.S. Department of Energy (DOE), *Hydropower Market Report 2023*, Office of Energy Efficiency and Renewable Energy, 2023. Available at: https://www.energy.gov/eere/water/hydropower-market-report
36. International Hydropower Association (IHA), *2024 Hydropower Status Report: Sector Trends and Insights*, London, 2024. Available at: https://www.hydropower.org/publications/2024-hydropower-status-report
37. ThinkGeoEnergy, *Global Geothermal Power Generation Statistics 2024*, Published May 2024.

Available at: https://www.thinkgeoenergy.com/global-geothermal-power-generation-2024/

38. ITER Organization, *ITER: The World's Largest Fusion Experiment*, Progress Reports and Updates, 2024. Available at: https://www.iter.org/news
39. United Nations Institute for Training and Research (UNITAR), *Global E-waste Monitor 2024*, in collaboration with ITU, ISWA, and UNU. Bonn/Geneva/Rotterdam, 2024. Available at: https://ewastemonitor.info
40. Indian Green Building Council (IGBC), *List of IGBC Platinum Certified Campuses and Buildings*, Confederation of Indian Industry (CII), 2023. Available at: https://igbc.in
41. Bureau of Energy Efficiency (BEE), *Energy Conservation Building Code (ECBC) 2017*, Ministry of Power, Government of India. Available at: https://beeindia.gov.in/documents/ECBC
42. Bureau of Indian Standards (BIS), *National Building Code of India (NBC) 2016*, BIS, New Delhi.
43. United Nations, *World Population Prospects 2022: Summary of Results*, Department of Economic and Social Affairs, Population Division, UN, 2022. Available at: https://www.un.org/development/desa/pd/content/world-population-prospects-2022
44. United Nations, *World Urbanization Prospects: The 2018 Revision*, Department of Economic and Social Affairs, UN, 2019. Available at: https://population.un.org/wup/
45. Food and Agriculture Organization of the United Nations (FAO), *Tackling Climate Change Through Livestock: A Global Assessment of Emissions and Mitigation Opportunities*, FAO, Rome, 2013. Available at: https://www.fao.org/3/i3437e/i3437e.pdf

46. Shell, *Shell Energy Scenarios to 2050: Future of Transport*, Shell International, 2023.
Available at: https://www.shell.com/energy-and-innovation/the-energy-future/scenarios.html
47. Battery University, *BU-205: Types of Lithium-ion Batteries*, Cadex Electronics Inc., 2024.
Available at: https://batteryuniversity.com/article/bu-205-types-of-lithium-ion
48. Shiva, Vandana. *Staying Alive: Women, Ecology, and Development*, Zed Books, London, 2016.
Also supported by:
Navdanya, *About the Movement*, Navdanya Official Website.
Available at: https://navdanya.org
49. Maathai, Wangari. *Unbowed: A Memoir*. Anchor Books, New York, 2006.
Also supported by:
The Green Belt Movement: *About Wangari Maathai*.
Available at: https://www.greenbeltmovement.org
50. Solar Energy Corporation of India (SECI), *Solar and Wind Energy Auction Results (2010–2024)*.
Available at: https://www.seci.co.in
51. The Guardian, *"How Fast Fashion is Hurting the Planet,"* citing Soil Association and Hubbub, 2019.

Additional Reading

- Green Technology for Sustainable Development by Ritu Singh and Sushil Kumar
- Renewable Energy Engineering and Technology by V.V.N. Kishore (TERI Press)
- Cradle to Cradle: Remaking the Way We Make Things by William McDonough and Michael Braungart
- India's Energy Transition by Ajay Mathur
- Environmental Management by N.K. Oberoi
- Renewable Energy Sources and Emerging Technologies by D.P. Kothari, K.C. Singal, Rakesh Ranjan
- Environmental Studies by Erach Bharucha
- Climate Change and Sustainable Development: A Holistic Approach by P.R. Trivedi
- Our Common Future (Brundtland Report by UN World Commission on Environment and Development
- Mission India: A Vision for Indian Youth by Dr. A.P.J. Abdul Kalam

Career in the Green Technology Sector

A. Job Opportunities

1. Government and Policy Roles

- Environmental Policy Analyst
- Renewable Energy Project Officer (MNRE, NITI Aayog)
- Sustainability Officer in Urban Local Bodies/Smart Cities Mission
- Green Auditor (MoEFCC, CPCB, SPCBs)
- Environmental Compliance Officer (Pollution Control Boards)
- In the central and state pollution control boards.

2. Private Sector and Corporate Roles

- Sustainability Consultant (ESG, CSR strategies)
- Renewable Energy Engineer (Solar, Wind, Biomass)
- Carbon Analyst / Climate Risk Analyst
- Green Construction and LEED Consultant
- Environmental Health and Safety (EHS) Manager

3. Research and Academia

- Research Scientist (Green Materials, Circular Economy)
- Assistant/Associate Professor (Environmental Engineering, Green Design)
- Project Associate (DST, DBT, UNDP-funded green projects)

4. Technical Roles

- Solar PV Designer / Installer / Maintenance Technician
- Smart Grid and Energy Efficiency Analyst
- EV Charging Infrastructure Technician
- Waste-to-Energy Plant Operator
- Green IT Specialist (Green Cloud, Eco-Software Engineering)

B. Business and Startup Opportunities

1. Renewable Energy
    - Solar Rooftop Installation Firm
    - Wind Microturbine Services for Rural Areas
    - Mini/Microgrid Provider for Remote Villages
    - Solar Cold Storage for Agriculture
2. Green Construction and Architecture
    - Green Building Materials Manufacturing (Fly-ash Bricks, Bamboo Panels)
    - LEED/BREEAM Certification Services
    - Sustainable Home Design Studio

3. Sustainable Transport
    - EV Charging Station Network
    - Battery Recycling and Second-life Battery Services
    - Electric Vehicle Dealership or Retrofit Kit Business

4. Waste Management and Circular Economy
    - e-Waste Collection and Refurbishing Units
    - Biodegradable Packaging Startup
    - Plastic Waste to Fuel/Brick Startup
    - Food Waste to Biogas Home Solutions

5. Water and Sanitation
    - Rainwater Harvesting System Manufacturer
    - Decentralized Wastewater Treatment Plant Provider
    - Water Purification Solutions using Solar Energy

6. Green Agriculture
    - Organic Fertilizer Production from Argo-waste
    - Vertical Farming Units
    - IoT-based Precision Agriculture Solutions
    - Aquaponics/Hydroponics Greenhouses

7. Carbon and ESG Services
    - Carbon Footprint Audit and Offset Service
    - Sustainable Supply Chain Verification
    - ESG Reporting Software Startup

C. Won't You Try These Emerging Startups?

- Climate Risk Modelling Platforms, Geoengineering R&D
- AI for Smart Grid Optimization, Blockchain for P2P Energy Trading
- Drone-based Crop Monitoring, Biochar Soil Enhancers
- Ocean Plastic Recycling, Tidal and Wave Energy Modules
- Bioplastics from Algae/Mushroom Mycelium, Green Nanomaterials
- Bioplastics from Banana leaves and Corn leaves
- Industrial Symbiosis Platforms (waste = input)

D. Support and Funding for Green Startups

- Startup India Green Sector Incentives
- Ministry of New and Renewable Energy (MNRE) subsidies
- Atal Innovation Mission (AIM) for cleantech incubation
- SIDBI Green Finance Scheme
- UNDP/UNIDO Clean Tech Innovation Programs
- International Solar Alliance (ISA) support

E. Entrepreneurial Training and Incubation Centres

- TERI – Technology Incubation for Sustainability
- IIT Madras – Incubation Cell for Renewable Startups
- Centre for Innovation Incubation and Entrepreneurship (CIIE) – IIM Ahmedabad
- NSRCEL – IIM Bangalore's Sustainability Cohort
- NIDHI-EIR and PRAYAS (DST Schemes)

Want to explore more?

Scan this QR code to witness melting glaciers, drying rivers, etc.	
Scan this QR code to see footage from NASA, ranging from wildfires to climate change.	
Scan this QR code to see India's largest floating solar power plant being commissioned.	

Hope you enjoyed the Green Voyage!

Index

www.ingramcontent.com/pod-product-compliance
Lightning Source LLC
LaVergne TN
LVHW090303210926
848480LV00001B/112